Sulfatases of Microbial Origin

Volume I

Authors

Kenneth S. Dodgson
Graham F. White
Department of Biochemistry
University College,
Cardiff, Wales, U.K.

John W. Fitzgerald
Department of Microbiology
University of Georgia
Athens, Georgia

CRC Press
Taylor & Francis Group
Boca Raton London New York

CRC Press is an imprint of the
Taylor & Francis Group, an **informa** business

First published 1982 by CRC Press
Taylor & Francis Group
6000 Broken Sound Parkway NW, Suite 300
Boca Raton, FL 33487-2742

Reissued 2018 by CRC Press

Library of Congress Cataloging in Publication Data

Dodgson, Kenneth S. (Kenneth Scott), 1922-
Sulfatases of microbial origin.

Bibliography: p.
Includes index.
1. Sulphatases. 2. Microbial enzymes.
I. White, Graham F. (Graham Francis),
1947- . II. Fitzgerald, John W. (John
Walter), 1938- . III. Title. [DNLM:
1. Sulfatases—Biosynthesis. 2. Sulfatases—
Physiology. 3. Microbiology. QU 136 D645s]
QP609.S95D63 576'.11925 81-15526
ISBN 0-8493-6035-8 (v. 1) AACR2
ISBN 0-8493-6036-6 (v. 2)

A Library of Congress record exists under LC control number: 81015526

Publisher's Note
The publisher has gone to great lengths to ensure the quality of this reprint but points out that some imperfections in the original copies may be apparent.

Disclaimer
The publisher has made every effort to trace copyright holders and welcomes correspondence from those they have been unable to contact.

ISBN 13: 978-1-315-89790-5 (hbk)
ISBN 13: 978-1-351-07700-2 (ebk)

Visit the Taylor & Francis Web site at http://www.taylorandfrancis.com and the
CRC Press Web site at http://www.crcpress.com

PREFACE

Review articles on the sulfatase enzymes have appeared at fairly regular intervals, but no book devoted exclusively to these enzymes has previously appeared. Information about the enzymes and their known and potential substrates is now so vast and impinges on so many different fields of scientific endeavor that the need for collated information has become critical. The present two volumes, although confined in their scope to those sulfatases that are present in microorganisms, will hopefully contribute towards meeting this need.

In the authors' experience, few investigators enter the sulfatase field because of their basic interest in the enzymes. More frequently, they have been concerned with other problems in areas such as drug metabolism, connective tissue biochemistry, or microbial genetics, and their involvement with sulfatase enzymes has often been incidental to the main objectives of their researches. This probably goes some way towards explaining why information about microbial sulfatases is so uneven in detail and extent, and why the results of some ventures into the field leave one with a feeling that it would have paid the investigators to have consulted the literature at the outset.

In the succeeding pages the authors have collated scattered information about individual sulfatases and have endeavored to indicate their physiological roles within the microorganisms, the ways in which their synthesis is subject to genetic and physiological control, and their participation in natural processes such as the recycling of sulfur. The authors have also attempted, for the first time, to discuss at length the mechanisms of action of some of the enzymes in relation to current knowledge about the nonenzymic hydrolysis of various types of sulfate ester. Although primarily directed towards those people interested in the biochemistry and enzymology of microorganisms, it is the authors' belief that there will be much in the book that will be of interest to workers in the mammalian field.

As a final point, the authors wish to thank the many people who have helped in one way or another with the birth of these volumes, and to pay particular tribute to the fortitude and stoicism exhibited by their respective wives during their literary preoccupations.

Kenneth S. Dodgson
Graham F. White
John W. Fitzgerald

THE AUTHORS

Kenneth S. Dodgson received the degrees of Ph.D. and D.Sc. from the University of Liverpool in 1949 and 1961, respectively. Since 1968 he has occupied the Chair of Biochemistry in University College, University of Wales, Cardiff and is currently acting as Vice Principal (Science) for that institution. He holds an honorary appointment as Visiting Professor to the University of Georgia and served as Honorary Meetings Secretary and Committee Secretary of the Biochemical Society during the period 1964 to 1970. He was a member of the British National Committee for Biochemistry from 1967 to 1971.

Professor Dodgson's research has been particularly concerned with the biochemistry and enzymology of sulfur-containing compounds. He is a Fellow of the Institute of Biology and a member of the Biochemical Society and Connective Tissue Society.

Graham F. White is presently a Lecturer in Biochemistry at University College, University of Wales, Cardiff. In 1968 he was awarded the degree of B.Sc. and the John Millar Thomson Medal for Chemistry at the University of London King's College. Postgraduate studies in enzyme kinetics at King's College led to the award of Ph.D. in 1971 whereupon he accepted a Royal Society/NATO Research Fellowship to pursue studies in enzymology at the Institute of Biological Chemistry, University of Florence, Italy.

In 1973 he joined the Department of Biochemistry in Cardiff as a Wellcome Trust Research Fellow to continue research on enzymes of sulfur metabolism, and was appointed to his present position in 1975. Dr. White is a member of the Royal Society of Chemistry.

John W. Fitzgerald is currently an Associate Professor in the Department of Microbiology at the University of Georgia. Dr. Fitzgerald obtained the BSc., MSc., and Ph.D. degrees with emphasis in Microbial Biochemistry from Queen's University, Kingston, Ontario, Canada. He left Queens in 1969 to take up a Leverhulme Postdoctoral Fellowship working with Professor K. S. Dodgson at University College, Cardiff. After serving as a Postdoctoral Associate of Professor W. J. Payne at the University of Georgia, Dr. Fitzgerald joined the faculty of the Department of Microbiology as an Assistant Professor in 1971.

TABLE OF CONTENTS

TABLE OF CONTENTS

Chapter 1

INTRODUCTION

The association between prestige and the color purple dates from pre-Roman times and reflects the early establishment of a flourishing dyestuffs industry in the Mediterranean region. So highly regarded was the Tyrian Purple produced by that industry that in the later days of the Roman Empire the wearing of a garment dyed with pure purple by a person who was not a member of the court could lead to sentence of death for treason.[1] The highly successful dyestuff industry depended, for its scientific basis, on the presence in various species of marine mollusk (genus *Murex)* of aryl sulfate esters with structures based on indoxyl, together with arylsulfatase enzymes that could use the esters as substrates. Removal of the sulfate group as a result of enzyme action was the initial stage in the transformation of the substrates to purple colored products related to indigotin. In 1911 this molluscan enzyme became the first of many sulfatases to be "discovered" in the laboratory.[2]

Sulfatases (sulfuric ester hydrolases, sulfohydrolases, EC 3.1.5) constitute a widely distributed class of enzymes that have the capacity to hydrolyse sulfuric acid esters according to the equation

$$R{-}O{-}SO_3^- + H_2O \rightarrow R{-}OH + SO_4^{2-} + H^+ \qquad (1)$$

where R can represent a remarkably diverse group of parent moieties and where the linkage with sulfate is in the form C–O–S. However, this account would be inadequate if it failed to include a number of analogous enzyme systems that act on substrates in which sulfate is present in linkage forms other than C–O–S. These substrates possess P–O–S linkages (e.g., adenylyl sulfate, generally termed adenosine 5′-phosphosulfate, APS), N–O–S linkages (as in the glucosinolates — mustard oil glycosides) or N–S linkages (e.g., the sulfamido groups of heparin), and are clearly not strictly sulfate esters nor substrates for true sulfatases. However, the relationships are so close and the compounds are so much a feature of the general biochemistry of sulfate that it is important and convenient to include them in these volumes. Accepting this wider definition, examples of sulfatase enzymes and potential substrates are to be found at all levels of complexity of living organisms and accounts of new examples are regularly appearing in the scientific literature.

I. POTENTIAL SUBSTRATES

The ability to synthesize sulfuric acid esters by transferring to suitable acceptors the sulfate group from 3′-phosphoadenylyl sulfate (generally referred to as 3′-phosphoadenosine 5′-phosphosulfate, PAPS) via appropriate sulfotransferase enzymes, is widely and sometimes curiously distributed in living organisms. For example, the influenza virus produces its own characteristic sulfated glycoprotein using the general mechanisms of the host cell,[3] but the significance of the product to the virus is quite obscure. Similarly, mammals are able to sulfate simple primary alcohols such as butan 2-ol,[4,5] yet neither the product nor alcoholic precursor appear to have physiological roles to play or are obvious features of normal mammalian metabolism. There is also a remarkable variation in the molecular complexity of naturally occurring sulfate esters, ranging from the simple methyl sulfate found in the hen's egg[6] to the polysulfated glycosaminoglycan-protein aggregates (proteoglycan aggregates) present in connective tissues.[7]

The practical consequences of esterifying with sulfuric acid a hydroxyl group of a suitable acceptor molecule are considerable. The ester sulfate group is profoundly acidic (pK well below that of the carboxyl group) and its introduction into a molecule frequently results in marked changes in the physical and other properties of a compound. Introduction of many such groups as in, for example, the polysulfated polysaccharides of seaweeds or the polysulfated glycosaminoglycans of connective tissues, produces polyanionic compounds that are able to bind counterions, to maintain a highly solvated domain and to adopt characteristic conformations, all properties that are presumed to be relevant to their roles within the living organism. The sulfate group inevitably confers increased water solubility and decreased lipid solubility on the parent molecule, both factors that are vital in the animal world to the excretion of phenolic and steroidal end-products of metabolism. In some instances the introduction of the sulfate group into a hydrophobic moiety leads to the appearance of surfactant properties, examples of which are to be found in the bile salts of many species of fish and amphibia. Other sulfate esters also possess surfactant properties (e.g., the various sulfolipids described in Volume I, Chapters 2 and 3) but there is, as yet, no obvious physiological advantage to the organism. Many of these compounds are present in membranous structures and their surfactant properties may be coincidental to the properties that determine their true role. Indeed there are indications that some of them may be involved in ion-transport phenomena.[8]

In a few instances the introduction of the sulfate group appears to mask hydroxyl groups that confer important physiological properties on a compound (e.g., estrogen sulfates) and it is possible that alternate sulfation and desulfation of those groups might constitute a mechanism for regulating those physiological properties. In contrast, in at least two cases the pharmacological properties of hormone-like polypeptides (phyllokinin and caerulein, present in the skins of certain amphibia,[9]) are greatly reduced by removal of the sulfate group from its parent tyrosine residue. There has also been a recent suggestion[10] that the sulfate groups of certain glycoproteins may have an important role to play in recognition phenomena either at cellular or molecular level. In one exceptional case, choline sulfate, the presence of the ester sulfate group actually allows the formation of an internally compensated, uncharged compound that serves as a readily utilizable sulfur store in a considerable number of fungi and (possibly) plants. Presumably, the internal compensation of charge allows large quantities to be stored with minimum disturbance of ionic or osmotic balance of the tissues. Of course, in many instances these attempts to assign physiological roles to sulfate esters are largely speculative, but the diversity and widespread distribution of the compounds is such as to indicate their importance to life, even though we have relatively little appreciation as yet as to what that importance might be.

So far the discussion has centered on naturally occurring sulfate esters, but it will be apparent from later sections of this book that chemically synthesized sulfate esters, sulfate ester precursors, and sulfamides (sulfamates) also enter the environment in one way or another. For example, the primary and secondary alkyl sulfate surfactants that are in common use as components of detergent formulations and which are returned to soils and waters in huge quantities. It will be seen in the following chapter that hitherto unsuspected sulfatase enzymes have been discovered as a result of the introduction of such compounds into the environment. However, it will also emerge that such enzymes almost certainly have other naturally occurring substrates with which they are normally concerned.

It will later become obvious from the contents of the present volumes that some sulfatases have been investigated in depth whereas others have been but sparingly studied. There are several reasons for this, but certainly an important one relates to the ease of preparation or commercial availability of the substrates or potential substrates. The preparation (and to a lesser degree, the commercial availability) of substrates for arylsulfatases, primary alkylsulfatases, and choline sulfatase presents no undue difficulties to the investigators. In contrast, the preparation of defined stereoisomers of secondary alkyl sulfate esters (substrates for stereospecific secondary alkylsulfatases) is fraught with difficulties. Similar considerations apply to potential substrates for carbohydrate sulfatases. Carbohydrate sulfates have generally been rather difficult to prepare or to isolate from natural sources and yet such compounds are among the most common of all sulfate esters. Additional complications arise because some carbohydrate sulfates are extremely complex, with chemical structures not yet adequately defined. Indeed, in some instances sulfatases have been discovered because of the desire to provide enzymes that might usefully be employed as tools in aiding structural determination. On occasions this approach to the elucidation of polymer structure has been most valuable and has pointed to the way in which other complex sulfated polymers might be studied in the future.

The assignment of appropriate chemical names to some of the sulfate esters that are mentioned in succeeding chapters is not helped by the structural complexity of the compounds in question and, in some cases, by an inadequate understanding of structure. An attempt has therefore been made to ease the burden of the nonspecialist reader by assigning trivial names that are frequently encountered in the literature or appear to the authors to convey a description of structure that will be meaningful to such a reader. In some cases where it has been deemed necessary, alternate names have been given or the correct chemical nomenclature has been cited at least once. With some of the esters that will be discussed the situation has been particularly confused by changes in nomenclature that have been made as further thoughts on structure or other factors have emerged; for example, the recent switch from "heparitin sulfate" to "heparan sulfate" by all active workers in the glycosaminoglycan (mucopolysaccharide) field (see Volume I, Chapter 3). The constant use in the literature of the abbreviations APS and PAPS (adenosine 5′-phosphosulfate and 3′-phosphoadenosine 5′-phosphosulfate, respectively) for compounds with the more recently recommended names of adenylyl sulfate and 3′-phosphoadenylyl sulfate, constitutes another case in point. In the ensuing text the authors have used heparan sulfate because of its general use in contemporary literature, but have retained the abbreviations APS and PAPS because of their continued widespread usage.

Strictly speaking, of course, a sulfate ester of the following structure, $C_6H_5 \cdot O \cdot SO_3H$, should be referred to as phenyl hydrogen sulfate rather than phenyl sulfate. Moreover, because sulfate esters generally exist in the form of salts under natural and experimental conditions, the compound under discussion should be named as a salt, for example, sodium phenyl sulfate. In accord with generally accepted practice the authors shall continue to use, for the most part, the shortest of the three alternatives.

II. THE ENZYMES

Turning now to the sulfatase enzymes it is worth recalling that they commanded relatively little attention until comparatively recently. This probably reflects the early difficulties experienced in accepting the enzymes as important when some of them seemed to have no obvious natural substrates or physiological roles. Moreover, some natural potential substrates appeared to have no corresponding enzymes. Nothing was more calculated to lead to loss of interest during periods when other aspects of enzymology were moving ahead at an exponential rate. Additionally, the measurement of activity of many sulfatase enzymes has always posed problems, some of which are still not adequately resolved. One end-product of sulfatase action is always inorganic sulfate, but the difficulties of measuring this on a microscale are considerable, particularly when it is present in crude preparations of biological material. In the Cardiff laboratories a number of different methods of approaching this problem have been made,[11-17] but none is entirely satisfactory and constant vigilance must be maintained to ensure that the method of choice is appropriate to the enzyme preparation and the substrates under examination. The newcomer to the field is recommended always to ensure that inorganic sulfate can be quantitatively recovered from enzyme-substrate mixtures under the precise experimental conditions to be used. Sulfate assay difficulties probably go some way to explaining why the arylsulfatase enzymes have generally been the major focus of attention of investigators. Chromogenic substrates (e.g., *p*-nitrophenyl sulfate) for the assay of arylsulfatases have been available for many years[18] and eliminate the need for the tedious and often unreliable determination of inorganic sulfate.

General disinterest in the sulfatase enzymes suddenly ceased when, following years of patient work and steady progress by a handful of workers, it became clear that a group of enzymes at one time considered to be evolutionary relics were of considerable importance to the physiology of animals and microorganisms. In the animal world certain sulfatases, long studied for their activity toward aryl sulfate esters, suddenly emerged as having a substrate specificity far wider than could have been anticipated. Thus, arylsulfatases A and B were shown to participate in the lysosomal degradation of cerebroside sulfates and glycosaminoglycan sulfates, respectively, substrates in which the susceptible sulfate groups were located on sugar residues. Perhaps the matter would have remained there had it not been for the fact that accumulation of the substrates of the two enzymes was a feature of metachromatic leucodystrophy and Maroteaux-Lamy syndrome, both fatal metabolic diseases in which substrate storage results from the deficiency of arylsulfatases A and B, respectively (see References 19-22 for reviews). An enormous quickening of interest followed and led to the discovery of new mammalian sulfatases that were also involved in the lysosomal degradation of sulfated glycosaminoglycans.

In the world of microorganisms, the quickening of interest can probably be traced back to three factors.

1. The discovery by Spencer's group in Dublin that the utilization of the choline sulfate stores of certain fungi was dependent on a sulfatase enzyme (choline sulfatase), the production of which was controlled by the sulfur status of the fungi (see Volume I, Chapter 2 and Volume II, Chapter 3).
2. The realization that microbial sulfatases and associated enzymes could serve as important aids to the study of structure and composition of complex glycosaminoglycan sulfates (see Volume I, Chapter 3).
3. The growing realization by soil microbiologists that vast quantities of sulfatase substrates were being constantly returned to the soils and waters of the earth, and that the sulfatases of soil and water microorganisms had a vital role to play in the recycling of sulfur (see Volume II, Chapter 4).

At the present time the sulfated compounds shown in Table 1 are known to serve as substrates for different microbial sulfatases. In column A are shown substrates of small molecular size, all of which can undergo desulfation without the need for the prior action of any other enzyme. Column B lists the polymeric substrates, all of which must first undergo depolymerization by endoglycosidases or endolyases before enzymic desulfation can occur. The most likely substrate that is presented during polymer biodegradation to each of the sulfatases involved is cited in parentheses. Alternative substrates are available for some of the sulfatases and these will be mentioned in the appropriate parts of the text. The possibility also exists that some of the compounds

Table 1
SOME SUBSTRATES FOR DIFFERENT MICROBIAL SULFATASES

A	B	
Adenosine 5′-phosphosulfate	κ-Carrageenan	(neocarrabiose 4-sulfate)
Aryl sulfates	Chondroitin 4-sulfate	(D-Δ4,5-glucuronopyranosyl [β1 → 3]-*N*-acetyl-D-galactosamine 4-sulfate)
Choline sulfate		
Primary alkyl sulfates	Chondroitin 6-sulfate	(D-Δ4,5-glucuronopyranosyl [β1 → 3]-*N*-acetyl-D-galactosamine 6-sulfate)
Secondary alkyl sulfates	Dermatan sulfate	(D-Δ4,5-glucuronopyranosyl [β1 → 3]-*N*-acetyl-D-galactosamine 4-sulfate)
2-(2,4-Dichlorophenoxy) ethyl sulfate	Heparan sulfate[a]	(L-Δ4,5-iduronopyranosyl [α1 → 3]-D-glucosamine 2,6-disulfate and L-Δ4,5-iduronopyranosyl[α1 → 4]-*N*-acetyl-D-glucosamine 6-sulfate)
D-Glucose 6-sulfate		
D-Gluconate 6-sulfate		
D-Glycerate 3-sulfate	Heparin[a]	(2-sulfo-L-Δ4,5-iduronopyranosyl [α1 → 4]-D-glucosamine 2,6-disulfate)
3′-Phosphoadenosine 5′-phosphosulfate	Keratan sulfate	(small oligosaccharide fragments containing *N*-acetyl-D-glucosamine 6-sulfate and D-galactose 6-sulfate residues)
Steroid sulfates		
Sulfamates		

Note: Compounds in column A are each desulfated by one sulfatase only and desulfation proceeds without the need for prior action of any other enzyme. Column B lists known polymeric substrates, each of which must first undergo depolymerization before desulfation. The major substrate(s) presented to the sulfatase(s) after depolymerization is cited in parentheses. More than one sulfatase may be involved.

[a] The same sulfatases are involved in the degradation of both polymers.

in column A may not reflect the true naturally occurring substrates of the sulfatases in question and this points again to a problem mentioned previously, namely that some of the enzymes do exhibit curious cross-specificities. Examples previously cited are mammalian arylsulfatases A and B, the true substrates of which possess sulfate groups on sugar residues. It is not impossible that something similar may eventually emerge in relation to certain microbial arylsulfatases (see Volume II, Chapter 4). In the mammalian world it has been shown that the lysosomal enzyme hydrolysing adenosine 5′-phosphosulfate (APS) will hydrolyse even more effectively compounds such as ATP, ADP, and pyrophosphate, and the enzyme is now believed to be of no significance to metabolic transformations involving APS (see Reference 23 for review and Volume II, Chapter 1). Particular care must now be taken, therefore, before assuming that a sulfatase shown to be active against a particular type of sulfate ester is actually acting on that type in a physiological situation. For example, it seems not unreasonable to ask whether certain fungal arylsulfatases present in lysosome-like granules (cf. mammalian arylsulfatases A and B) really act on aryl sulfate esters in a physiological situation, a point that will be raised again in Volumes I and II, Chapter 4.

At the present time our knowledge of microbial sulfatases is very limited and the whole field is ripe for investigation. There are a number of reasons why these enzymes can prove to be attractive and compelling subjects of study for enzymologist and microbiologist alike. In the first place, a number of sulfatases exhibit properties that make them particularly fascinating to the enzymologist interested in reaction mechanisms. For example, studies on alkylsulfatases have revealed a class of enzymes unique among simple ester hydrolases in that they catalyze the scission of the oxygen– alcohol bond of the ester linkage, rather than the usual oxygen– acid bond. Moreover, those enzymes that act on secondary alkyl sulfates exhibit stereospecificity for D- or L- forms, with bond scission being accompanied by bond inversion (see Volumes I and II, Chapter 2). In early studies on arylsulfatases numerous unsuccessful attempts were made to test the possibility that the enzymes could act as transferases. Recently, it has become clear that one of three arylsulfatases present in *Aspergillus oryzae* possesses the ability to transfer sulfate from a substrate such as *p*-nitrophenyl sulfate to a suitable amine acceptor such as tyramine. The relationship between the three enzymes in terms of mechanism of action and substrate binding and catalytic sites presents a fascinating problem that awaits investigation. A further intriguing problem relates to the fact that some sulfatases are strongly inhibited by sulfate ions whereas others are not so affected. This property difference has long been used to distinguish two distinct types of arylsulfatase.[18] A possible explanation for the difference will be offered in Volume II, Chapter 2, but experimental evidence in its favor has yet to be sought. Thus far the enzyme technologist has not been involved with sulfatases to any significant extent, but the commercial availability and successful use of certain sulfatase enzymes as research tools in structural studies of complex sulfated polymers suggests that this may well be a future growth area.

The microbial physiologist will find plenty to occupy his attention in the field. For example, relatively little is known about the type of transport mechanisms that operate for the different sulfatase substrates. Similar considerations apply to the cellular localization of the enzymes, particularly in those cases where additional enzymes other than sulfatases are also involved in the total biodegradation of a polymeric sulfate ester. The geneticist is also presented with an abundance of opportunities. The induction, regulation of induction, presence of multiple forms of the same enzyme and different types of sulfatases within the same microorganism, and genetic loci in relation to the genes of other enzymes concerned with sulfur metabolism are obvious targets for research. Soil microbiologists have made only token incursions into the study of the biodegradation of the majority of known sulfate esters, the types of microorganisms involved, and the relationships of the organisms one to another. Many of the soils of the world are already sulfur deficient,[24] and yet efforts to understand the roles played by microorganisms in the recycling of ester sulfate sulfur are extremely limited in their scope and in the number of scientists involved.

In succeeding pages the authors have tried to draw attention to the deficiencies in our knowledge and to the opportunities that invite attention. Other similar deficiencies and opportunities will no doubt occur to the reader.

Finally, in some respects the advancement of knowledge concerning the sulfatases has proceeded relatively slowly, but in relation to the nomenclature of the enzymes discovery has tended to outstrip the assignment of adequate names. Thus, the latest edition of the recommendations of the I.U.B. Commission on Enzyme Nomenclature makes no reference to the microbial sulfatases that remove sulfate from lithocholate sulfate, κ-carrageenan, primary and secondary alkyl sulfates, and keratan sulfates. This in spite of the fact that some of these enzymes have been under investigation for a decade or more. Other instances will be cited as they become relevant. In the text that follows the authors shall sometimes use names that appear to be at variance with those recommended by the commission but, in each case, the recommended name will be cited at some stage. Some years are likely to elapse before enough is known about the specificities of the enzymes to allow a fully systematic and meaningful nomenclature to be adopted. Meanwhile, the enzymes themselves are profoundly more interesting than the names that may have been assigned to them.

REFERENCES

1. **Baker, J. T.**, Tyrian purple: an ancient dye, a modern problem, *Endeavour*, 33, 11, 1974.
2. **Derrien, M.**, Meeting of the Société Chimique de France, Montpelier, 16 December 1910, *Bull. Soc. Chim. Biol.*, 9, 110, 1911.
3. **Nakamura, K. and Compans, R. W.**, The cellular site of sulfation of influenza virus glycoprotein, *Virology*, 79, 381, 1977.
4. **Vestermark, A. and Boström, H.**, On the enzymatic formation of ester sulfates of primary aliphatic alcohols, *Exp. Cell. Res.*, 18, 174, 1959.
5. **Spencer, B.**, Endogenous sulphate acceptors in rat liver, *Biochem. J.*, 77, 294, 1960.
6. **Yagi, T.**, Sulfate esters in developing eggs, *J. Biochem. Tokyo*, 59, 495, 1966.
7. **Rodén, L. and Horowitz, M. I.**, Proteoglycans and structural glycoproteins, in *The Glycoconjugates*, Vol. 2, Horowitz, M. I. and Pigman, W., Eds., Academic Press, New York, 1977, 3.
8. **Karlsson, A., Samuelson, B. E., and Steen, G. O.**, Lipid composition and Na^+-K^+-dependent adenosine triphosphatase activity of the salt(nasal) gland of eider duck and herring gull: a role for sulphatides in sodium-ion transport, *Eur. J. Biochem.*, 46, 246, 1974.
9. **Dodgson, K. S. and Rose, F. A.**, Sulfoconjugation and sulfohydrolysis, in *Metabolic Conjugation and Metabolic Hydrolysis*, Vol. 1, Fishman, W. H., Ed., Academic Press, New York, 1977, 239.
10. **Kent, P. W., Coles, C. J., Cooper, J. R., and Mian, N. R.**, Sulfate ester groups as potential informational regulators in glycoproteins, in *Carbohydrate Sulfates*, Schweiger, R. G., Ed., American Chemical Society, Washington, D.C., 1978, chap. 3.
11. **Dodgson, K. S. and Spencer, B.**, The determination of inorganic sulphate in the study of sulphatases, *Biochem. J.*, 55, 436, 1953.
12. **Lloyd, A. G.**, The use of barium chloranilate in the determination of enzymically liberated sulphate, *Biochem. J.*, 72, 133, 1959.
13. **Spencer, B.**, The ultramicro determination of inorganic sulphate, *Biochem. J.*, 75, 435, 1960.
14. **Dodgson, K. S.**, Determination of inorganic sulphate in studies on the enzymic and nonenzymic hydrolysis of carbohydrate and other sulphate esters, *Biochem. J.*, 78, 312, 1961.
15. **Dodgson, K. S. and Tudball, N.**, Enzymic desulphation of L-serine *O*-{^{35}S}-sulphate: the intracellular localization of the enzyme, *Biochem. J.*, 81, 68, 1961.
16. **Denner, W. H. B. and Stokes, A. M.**, Rapid chromatographic method for the separation of nucleotide sulphates, *J. Chromatog.*, 69, 426, 1972.
17. **Cloves, J. M., Dodgson, K. S., White, G. F., and Fitzgerald, J. W.**, Purification and properties of the P2 primary alkylsulphohydrolase of the detergent-degrading bacterium *Pseudomonas* C12B, *Biochem. J.*, 185, 23, 1980.
18. **Dodgson, K. S. and Spencer, B.**, Assay of sulphatases, *Meth. Biochem. Anal.*, 4, 211, 1957.
19. **Dodgson, K. S. and Rose, F. A.**, Sulfohydrolases, in *Metabolism of Sulfur Compounds*, Greenberg, D., Ed., Academic Press, New York, 1975, chap. 9.
20. **Roy, A. B.**, Sulfatases, lysosomes and disease, *Aust. J. Expt. Biol. Med. Sci.*, 54, 111, 1976.
21. **Roy, A. B.**, Sulfatase A — an arylsulphatase and a glycosulphatase, in *Sulphur in Biology*, Ciba Foundation Symposium, No. 72, Excerpta Medica, Amsterdam, 1980, 177.
22. **Farooqui, A. A.**, Sulfatases, sulphate esters and their metabolic disorders, *Clin. Chim. Acta*, 100, 285, 1980.
23. **Dodgson, K. S. and Rose, F. A.**, Observations on the biological roles of sulphatases, in *Sulphur in Biology*, Ciba Foundation Symposium, No. 72, Excerpta Medica, Amsterdam, 1980, 163.
24. **Kelly, D. P.**, The sulphur cycle, in *Sulphur in Biology*, Ciba Foundation Symposium, No. 72, Excerpta Medica, Amsterdam, 1980, 3.

Chapter 2

THE ALKYLSULFATASES

I. INTRODUCTION

A somewhat heterogeneous and as yet ill defined group of enzymes can be conveniently described under this general heading and the group can then be subdivided as follows:

1. Simple alkylsulfatases (not listed by the Enzyme Nomenclature Committee of the I.U.B.) acting on substrates in which a primary or secondary alcohol hydroxyl group present in an alkyl chain is esterified with sulfate
2. Choline sulfatase (choline sulfate sulfohydrolase, EC 3.1.6.4) exhibiting a relatively high specificity for choline sulfate
3. Phenoxyethyl sulfatase (not listed by the Enzyme Nomenclature Committee) acting on 2-(2,4-dichlorophenoxy)ethyl sulfate (CRAG herbicide)
4. Sterol sulfatase (sterol-sulfate sulfohydrolase, EC 3.1.6.2) acting on ester sulfate groups attached to the saturated sterol A ring

The ability of certain pseudomonads to liberate sulfate from L-serine *O*-sulfate and L-threonine *O*-sulfate[1] might at first sight be thought to reflect the activity of an alkylsulfatase. However, detailed studies[2,3] on a soil bacterium designated as *Pseudomonas* FR revealed the presence of two pyridoxal phosphate dependent enzymes (one constitutive and one inducible) which degraded L-serine *O*-sulfate to pyruvate, ammonia, and sulfate by an $\alpha\beta$-elimination mechanism. A similar but pyridoxal phosphate-independent enzyme, L-serine *O*-sulfate ammonia lyase, (EC 4.3.1.10) had previously been demonstrated in mammalian tissues.[4,5] It is also not without significance that mammalian aspartate aminotransferase and alanine aminotransferase can also degrade L-serine *O*-sulfate via a pyridoxal phosphate dependent $\alpha\beta$-elimination mechanism. Because the sulfated amino acid does not occur naturally, there can be little doubt that both bacterial and mammalian enzymes that degrade it with release of sulfate are normally concerned with other substrates.

II. SIMPLE ALKYLSULFATASES

Microbial enzymes of this type have recently been receiving increasing attention because of the part that they play in the biodegradation of detergent preparations containing long-chain primary or secondary alkyl sulfate esters. For reasons that will appear later, they have proved to be perhaps the most novel of all the microbial sulfatases yet discovered.

Detergent formulations containing primary or secondary alkyl sulfates have now been in use for almost half a century and still constitute a significant proportion of the commercial market. As a consequence of their domestic and industrial use they are returned in considerable quantities to our soils and waters where they are fairly rapidly degraded by microorganisms. It seems clear (see later) that the initial stage in the biodegradation process is the enzymic removal of the ester sulfate group as a result of the action of primary or secondary alkylsulfatase enzymes.

Of course, the sulfated detergent preparations and the defined individual primary and secondary alkyl sulfates (e.g., dodecyl sulfate, octan-2-yl sulfate) that have been used in enzymological studies are synthetic compounds. However, it is not generally appreciated that naturally occurring alkyl sulfate esters are constantly being returned to soils and waters and may well serve as substrates for these simple primary and secondary alkylsulfatases or for closely related enzymes still awaiting discovery. Such potential substrates include the following:

1. The chlorinated and nonchlorinated alkyl diol disulfates, based on docosane or tetracosane and found in *Ochromonas danica* and other unicellular algae. The structures of several such compounds have so far been described with certainty[6] (see Figure 1, I-VIII), but others exist and it is probable that this type of compound occurs more widely than would at present be apparent.[7] It has already been shown[8] that a closely related synthetic compound, octadecan 1,12-disulfate, can serve as a substrate for both primary and secondary alkylsulfatases of *Pseudomonas* C12B.
2. The C_{18}, C_{20}, and C_{30} alkyl sulfates and alkyl disulfates of the brown seaweeds (*Fucus* sp.). The compounds shown in Figure 1, (IX-XII) have been described[9] and others may well exist.
3. The 2,3-dibromo-5-hydroxybenzyl-1′-4-disulfate of the red alga, *Rhodomela larix*,[10] and containing both alkyl- and aryl sulfate ester groups (Figure 2).
4. Cardiospermine 5-sulfate,[11] a sulfated, cyanogenic, β-D-linked glycoside present in *Cardiospermum grandiflorum* (Figure 2).
5. The bile alcohol sulfates present in many fish and amphibian biles. Haslewood[12] has provided an extensive list of organisms known to produce such compounds. However, it can confidently be predicted that many other sources remain to be discovered. Essentially, the compounds possess a steroid nucleus to which is attached an alkyl chain carrying an ester sulfate group on either a primary or secondary alcohol hydroxyl group (Figure 3). In at least one case, myxinol disulfate, a second sulfate group is present on the 3β-hydroxyl of the steroid A-ring.
6. The sulfate esters of methanol and propan 2-ol present in the developing hen's egg.[13] Monosulfate esters of propan 1,2-diol and butan 1,2-diol are also found in the egg, but the precise location of sulfate on these compounds appears not to have been determined. It seems likely that the distribution of such esters will not be restricted to the domestic fowl. Relating to the occurrence of these short-chain alkyl sulfates, it is also worth noting that methyl sulfate has recently been identified as a significant component of airborne particulate matter resulting from coal combustion processes.[14] Presumably this material eventually finds its way into the soils and waters.
7. The C_{20} and C_{21} sulfate ester end-products of steroid metabolism present in mammalian urine, plasma, and some tissues.[15,16] For example, cortisol 21-sulfate and pregn-5-ene-3β-20α-diol disulfate contain, respectively, sulfated primary and secondary alcohol groups that might, in theory, serve as potential alkylsulfatase substrates. (Figure 4).

A. Discovery of the Enzymes

An early pointer to the existence of simple alkylsulfatases came as a result of the chance observation of bacterial contamination in a commercial shampoo preparation, and the subsequent demonstration that certain pseudomonads grew well on solid media containing dodecyl sulfate.[17] However, it was left to the work of Payne and Feisal[18] and Hsu[19] to provide the real clue to the presence of such enzymes in living organisms. The former workers collected soils from the vicinity of a sewage outfall and isolated from them two Gram-negative bacteria (designated C12 and C12B) that were able to grow on synthetic media-containing sodium dodecyl sulfate as sole source of carbon. The foaming capacity of the detergent was destroyed by both organisms. Only one of

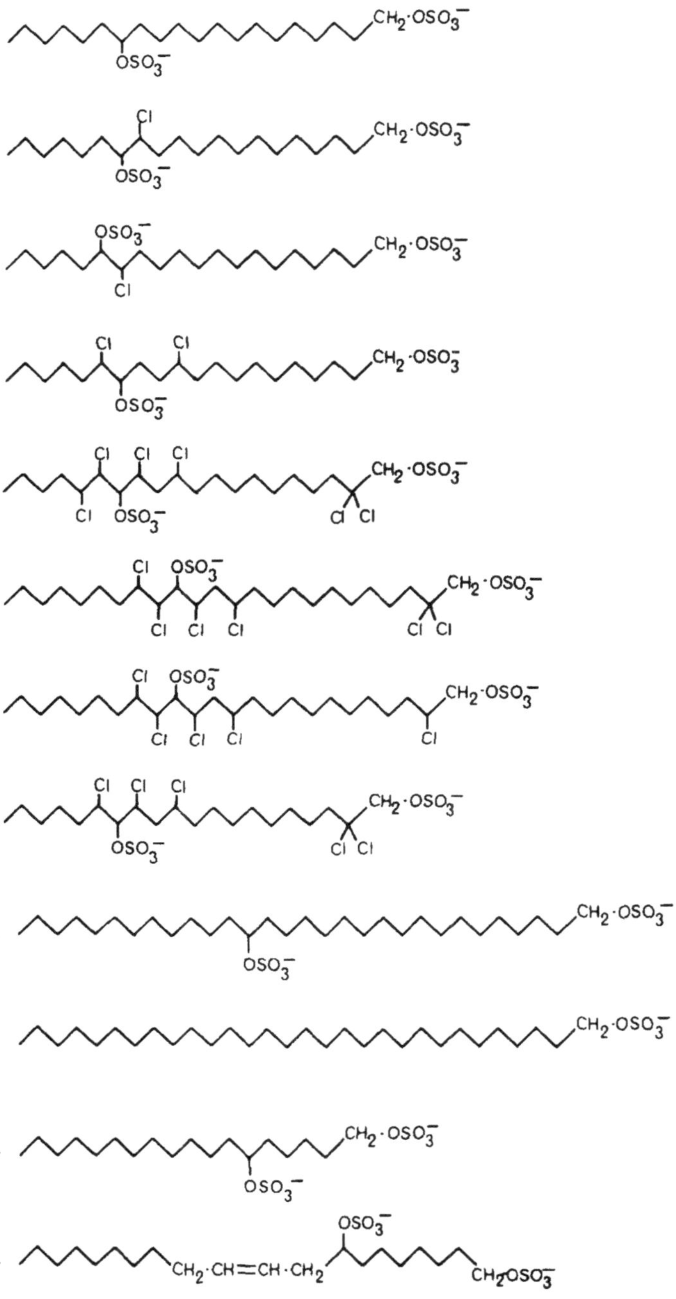

FIGURE 1. Naturally occurring examples of straight chain alkyl sulfates that might be potential substrates for simple alkylsulfatases. I, Docosanyl 1, 14-disulfate; II, 13-chlorodocosanyl 1, 14-disulfate; III, 14-chlorodocosanyl 1,15-disulfate; IV, 11,15- dichlorodocosanyl 1,14-disulfate; V, 2,2,11,13,15,16-hexachlorodocosanyl 1,14-disulfate; VI, 2,2,12,14,16,17-hexachlorotetracosanyl 1,15-disulfate; VII, 2,12,14,16,17-pentachlorotetracosanyl 1,15-disulfate; VIII, 2,2,11,13,15-pentachlorodocosanyl 1,14-disulfate; IX, triacontanyl 1,18-disulfate; X, triacontanyl sulfate; XI, octadecyl 1,6-disulfate; XII, eicos-10-enyl 1,8-disulfate.

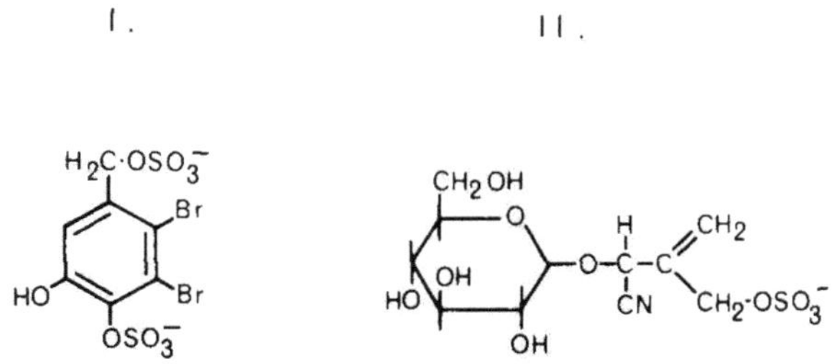

FIGURE 2. I, 2,3-Dibromo-5-hydroxybenzyl-1,4-disulfate from *Rhodomela larix;* II, cardiospermine 5-sulfate from *Cardiospermum grandiflorum.*

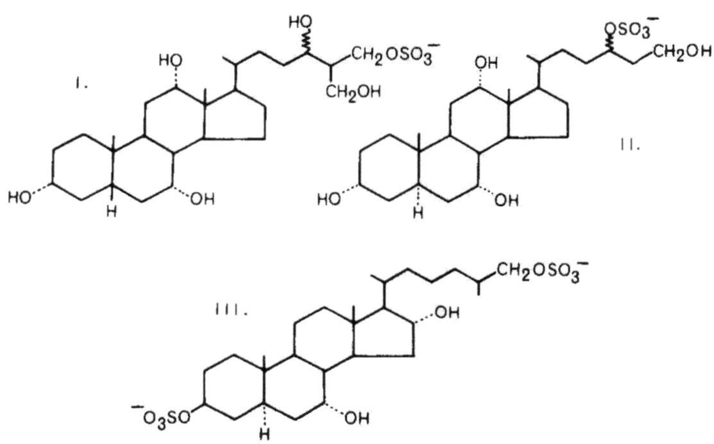

FIGURE 3. Examples of bile alcohol sulfates. I, scymnol sulfate; II, ranol sulfate; III, myxinol disulfate.

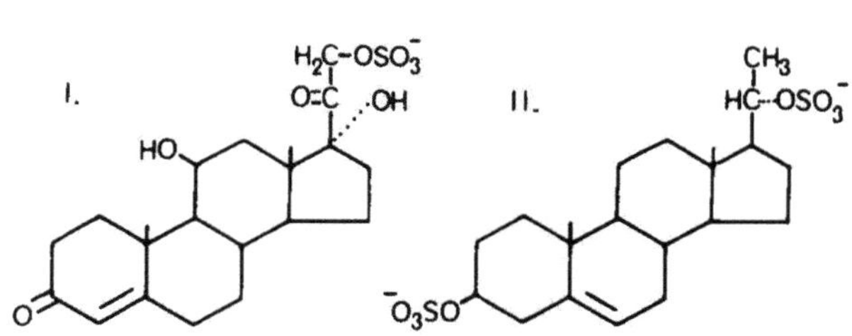

FIGURE 4. Examples of sulfated steroids that might have the potential to serve as substrates for simple alkylsulfatases. I, cortisol 21-sulfate; II, pregn-5-ene-3β,20α-diol disulfate.

the bacteria (C12B) was capable of growth when a detergent preparation containing alkyl benzene sulfonates was substituted as sole carbon source and it is this particular organism (now designated as *Pseudomonas* C12B) that has subsequently provided most of the detailed information about simple alkylsulfatase enzymes.

Other parallel studies from Payne's laboratory[20] established that resting cells of *Pseudomonas* C12B were able to oxidize straight-chain primary alcohols, including dodecanol. Activity toward the latter was markedly increased if the organism was grown on nutrient broth containing 0.1% sodium dodecyl sulfate, and an NAD-dependent dehydrogenase, active toward dodecanol, could be detected in extracts of cells grown in this way. It was suggested that an early event in the biodegradation of dodecyl sulfate was loss of the sulfate group and this was followed by oxidation of the liberated alcohol. Support for this view came later from the demonstration[21] that extracts of cells grown in the presence of dodecyl sulfate were able to liberate inorganic ^{35}S-sulfate from ^{35}S-labeled dodecyl sulfate.

Contemporary with this work was that of Hsu[19,22] who isolated from raw sewage water several undefined pseudomonads that were able to utilize dodecyl sulfate as carbon and sulfur sources for growth. Extracts of the cells of one of the organisms that had been adapted to dodecyl sulfate could liberate inorganic sulfate from that ester, as well as from octyl-, decyl- and tetradecyl sulfates. The activity of the intact cells was such that about 1.8 nmol of dodecyl sulfate was hydrolysed per 10^6 cells per hour. The branched-chain secondary alkyl sulfate, 2-methyl-6-ethylundecan 4-yl sulfate, could not serve as carbon and sulfur source for growth. However, because of the relative resistance to biodegradation of branched-chain detergents, this should not be taken to mean that the organism was unable to produce a secondary alkylsulfatase.

Hsu appeared to abandon his interesting studies at this point, but Payne and his colleagues continued to build up a general picture of simple alkylsulfatase activity in *Pseudomonas* C12B. It was shown[23] that an ammonium sulfate precipitated fraction from cell-free extracts liberated approximately equimolar amounts of inorganic ^{35}S-sulfate and dodecanol from ^{35}S-labeled dodecyl sulfate, thereby apparently confirming the enzymic reaction to be a simple hydrolysis of an ester bond. Arylsulfate esters were not hydrolysed by the fraction, neither was 2-(2,4-dichlorophenoxy)ethyl sulfate (CRAG herbicide) nor a detergent mixture of C_{10}-C_{20} secondary alkyl sulfates (hereafter referred to as Oronite®, and obtained from the Oronite division of the Chevron Chemical Co., San Francisco). Further studies[24] on the metabolism of a wide variety of odd and even numbered linear and branched primary alcohols and linear secondary alcohols confirmed that cell extracts contained one or more NAD-linked dehydrogenase enzymes active toward these alcohols. It was later concluded[25] that additional enzyme (rather than a new enzyme) was synthesized when the organism was grown on either dodecyl sulfate or a mixture of long-chain primary alcohols. In any event, the collective observations gave further support to the view that the biodegradation of a primary alkyl sulfate detergent was initiated by a primary alkylsulfatase, and that an NAD-linked alcohol dehydrogenase then converted the liberated alcohol to the corresponding carboxylic acid which could then serve as a substrate for β-oxidation.

In 1967, a new twist to these studies of Payne and his colleagues was provided by the isolation of a microorganism from soil enriched with Oronite®.[26] The organism was tentatively identified as *Aerobacter cloacae* and extracts prepared from cells grown on a mineral salts medium containing 0.25% Oronite® as carbon source were not only able to liberate inorganic sulfate from dodecyl sulfate, but also from "pentan 3-yl sulfate" (quotation marks have been appended to this secondary alkyl sulfate ester because it is now known to contain isomeric impurities: similar considerations apply to all other secondary alkyl sulfates that have been prepared by sulfating secondary alcohols with sulfuric acid or chlorosulfonic acid. Quotation marks will be used in these cases also).

Because extracts of dodecyl sulfate adapted cells of *Pseudomonas* C12B were devoid of sulfatase activity toward Oronite®, it seemed clear that a new type of alkylsulfatase enzyme, active toward secondary alkyl sulfates, had been discovered. It is interesting (and fortuitous in the light of subsequent developments) that the Payne group did not persist with *Aerobacter cloacae* as the source of this new enzyme, but turned again to *Pseudomonas* C12B in a search for a similar enzyme. The search was greatly facilitated by the development[25,27] of an ingenious method for detecting primary and secondary alkylsulfatase enzymes after polyacrylamide gel electrophoresis. Polyacrylamide gels, run under nondenaturing conditions, are extruded from the gel tubes and incubated in buffered solutions of potential alkyl sulfate substrates of chain length C_9 or greater. Where enzymes acting on the particular substrate used are present in the gel, a readily visible white band is formed consisting of microdroplets of liberated insoluble alcohol.

With this method, up to three separate alkylsulfatase bands were detected in extracts of cells of *Pseudomonas* C12B grown under different conditions,[25] Dodecyl sulfate was employed as the substrate for primary alkylsulfatase activity and "decan 5-yl sulfate" (now known to be contaminated with isomers, particularly the corresponding 2-yl sulfate) as substrate for secondary alkylsulfatase activity. When cells were grown to the stationary phase on nutrient broth alone, a single band of activity was observed, irrespective of which of the two substrates was used. If dodecyl sulfate was included in the broth an additional sulfatase band appeared, irrespective of the substrate used. Finally, if broth containing Oronite® was the growth medium, a third band also appeared if "decan 5-yl sulfate" was the substrate. It was concluded that *Pseudomonas* C12B could synthesize a total of three alkylsulfatases, two of them possessing dual specificity toward both primary and secondary alkyl sulfates and the third being active toward secondary alkyl sulfates only. In fact, these conclusions were oversimplified and other experiments[28] in which a variety of short-chain alkyl sulfates were tested as possible enzyme inducers lent support to this view.

It was at this point that collaboration between the microbiologists in the University of Georgia and the biochemists in University College Cardiff began. Refinement of the gel electrophoresis procedure soon led to the discovery[29] that each of the two gel bands with activity toward both primary and secondary alkyl sulfates actually contained distinct primary and secondary alkylsulfatases. In short, *Pseudomonas* C12B, under appropriate culture conditions, could produce up to five different alkylsulfatases. When the bacterium was grown on nutrient broth alone, an enzyme (designated P1) active toward primary alkyl sulfates, and two others (designated S1 and S2) that hydrolyse only secondary alkyl sulfates, could subsequently be detected on gels. These three enzymes appear to be constitutive. The inclusion of a primary alkyl sulfate in the nutrient broth caused the induction of a new primary alkylsulfatase (designated P2), so that a total of four enzyme bands then appeared on gels. Finally, the substitution of Oronite® for the primary alkyl sulfate in the broth resulted in the induction of a new secondary alkylsulfatase (S3), and five enzyme bands were detected on gels. The positions of these various bands on the gels are shown in Figure 5.

Two further comments might now be made. First, it is clear from later work[30] that the P2 primary alkylsulfatase has a somewhat transient existence and maximum specific enzyme activity is attained during the exponential phase of growth. This situation contrasts with that of the other four enzymes which all attain maximum specific activity in the early stationary phase. Second, fractionation of the Oronite® preparation used to induce S3 revealed[29] a complex situation, as two components, a sulfate ester fraction and a free alcohol fraction, were both needed for induction to occur. It was further shown that any combination of ester and alcohol from "tetradecan 2-yl sulfate", "hexadecan 2-yl sulfate", tetradecan 2-ol and hexadecan 2-ol could replace Oronite® as inducer. A more detailed account of the induction of the S3 and P2 enzymes can be found in Volume II, Chapter 3.

At this stage of progress it was clear that a major question needed to be resolved, namely, why *Pseudomonas* C12B should have the potential to produce three secondary alkylsulfatases and two primary alkylsulfatases. The answer to this question would clearly require the separation and possibly the purification of the individual enzymes in order to establish whether important fundamental differences existed in their specificities or other properties.

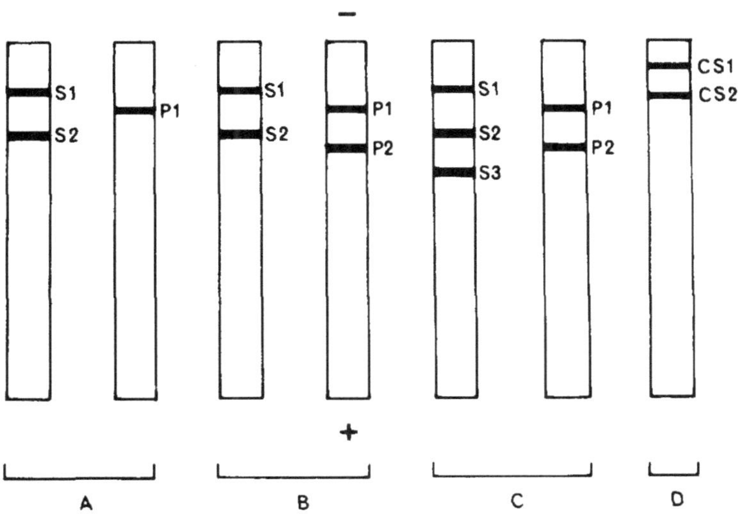

FIGURE 5. Idealized diagram showing relative electrophoretic mobility of primary and secondary alkylsulfatases of *Pseudomonas* C12B and *Comamonas terrigena* on polyacrylamide gels. *Pseudomonas* C12B grown on: A, nutrient broth; B, nutrient broth plus dodecylsulfate; C, nutrient broth plus Oronite®. *C. terrigena* grown on: D, nutrient broth. The letters S and P refer to secondary and primary enzymes, respectively.[25,27,29]

Mention should now be made of a microorganism that was coisolated with *Pseudomonas* C12B, but not examined further at the time. This organism, identified as a strain of *Comamonas terrigena,* now became a focus of attention because of its ability to produce two alkylsulfatases only, irrespective of the culture conditions employed.[31] Both enzymes (designated CS1 and CS2, see Figure 5) were specific for secondary alkyl sulfate ester substrates. Clearly, the chances of separating these two enzymes for further study were at least equally as great as the chances of separating the five enzymes of *Pseudomonas* C12B, and the purification of enzymes from both organisms was therefore initiated. In the event, the task proved to be less formidable in the case of the *C. terrigena* enzymes, one of which (CS2) was separated from the other and purified to homogeneity.[32] Studies on this enzyme were to provide the clue to the reason for the multiplicity of secondary alkylsulfatases in both microorganisms and led to the eventual realization that simple alkylsulfatases were indeed unique in relation to other sulfatase enzymes.

B. The CS2 Secondary Alkylsulfatase of *Comamonas terrigena*

The designation of this enzyme as CS2 was intended to reflect the type of organism *(Comamonas),* the enzyme specificity (secondary alkylsulfatase), and the fact that the enzyme moves toward the anode at a faster rate than the coexisting secondary alkylsulfatase (CS1). In the event, the designation could serve as a focus for confusion because, as will become apparent from later sections, the CS2 enzyme is basically similar in specificity to the S1 alkylsulfatase of *Pseudomonas* C12B.

It is also important to mention that the preliminary studies on the two *Comamonas* enzymes was carried out using impure substrates, namely, "octan 2-yl sulphate", "decan 2-yl sulfate", and "decan 5-yl sulfate". The investigators were unaware of this problem at that time.

1. Purification and General Properties

The organism was maintained on nutrient agar slants and was transferred monthly, and has so been maintained without apparent change in characteristics and properties for several years. It was grown on a large scale at 30° C in 350 ℓ of aerated nutrient broth, care being taken to harvest cells over the period of greatest enzyme activity. Preliminary studies[31] had indicated that the production of the CS1 and CS2 enzymes began at the onset of the stationary phase of growth, with maximum activity being reached in the late stationary phase immediately before significant cell lysis occurred. There was no problem associated with the storage of freeze-dried ruptured cells at −20° C and enzyme activity was retained for several months under these circumstances.

The enzyme purification procedure[32] involved the preparation of a cell-free extract, removal of nucleic acid on a DEAE-cellulose column, fractionation with ammonium sulfate, gel filtration on Sephadex® G-100 and fractionation on DEAE-cellulose. Appreciable losses of the coexisting CS1 enzyme occurred at the various stages, the last traces being eliminated in the final stage. The purification achieved was of the order of 4000-fold with a retention of 37% of the original activity.

The final preparation moved as a single band on polyacrylamide gel electrophoresis and as a single peak during gel filtration. Molecular weights of 250,000 and 249,000 were obtained by gel filtration on Sephadex® G-200 and Sepharose® CL-6B, respectively. A single band of molecular weight equivalent to 57,000 was given by SDS/polyacrylamide gel electrophoresis, indicating that the intact protein was probably a tetramer.

Initial studies on the pure enzyme using "octan 2-yl sulfate" as assay substrate indicated apparently suitable experimental conditions for kinetic studies over short incubation periods with less than 10% of the substrate being hydrolysed. However it soon became apparent, when incubation of enzyme and substrate was prolonged, that enzyme activity ceased fairly abruptly when about 40% of the substrate had disappeared.[33] Addition of further enzyme at this point was without effect, but led to the hydrolysis of 40% of the amount added. In contrast, crude cell extracts were able to hydrolyse about 80% of the substrate. Two important findings emerged from a detailed study of these phenomena. First the CS2 enzyme was shown to be stereospecific for secondary alkyl sulfates with a D-configuration (see also Reference 34) and second, that "octan 2-yl sulfate" and other secondary alkyl sulfates prepared by sulfation of secondary alcohols with sulfuric or chlorosulfonic acids, contained enzyme resistant isomeric impurities in addition to the expected D- and L- components. Interestingly, a close examination of the products of the enzymic or acid hydrolysis of "decan 5-yl sulfate" showed that the preparation contained more decan 2-yl sulfate than the expected 5-yl sulfate. It will later become apparent that the decision to employ "decan 5-yl sulfate" as substrate for developing secondary alkylsulfatase bands on gel electrophoretograms, was particularly fortunate. Had either a pure alkan 2-yl sulfate or a pure alkan 5-yl sulfate been employed then, respectively, either the S3 or the S1 and S2 enzymes might not have been discovered.

The development of milder sulfation techniques[33] subsequently enabled pure samples of D- and L-forms of octan 2-yl sulfate, as well as other racemic secondary alkyl sulfate esters, to be prepared. Tests of these various compounds as potential substrates for the CS2 secondary alkylsulfatase confirmed the stereospecificity requirement for a D-secondary alkyl sulfate and established a further specificity requirement, namely, that the sulfate group must be present at C-2 of the alkyl chain. When racemic samples of alkan 2-yl sulfates of increasing chain length were tested as substrates, it became clear that the enzyme had little or no activity toward substrates of chain length C_5 or less, and that activity increased as chain length increased from C_6 to C_{10}. Methodology difficulties prevented substrates with longer chains from being studied quantitatively at that period, but gel electrophoresis experiments confirmed that the enzyme could also hydrolyse racemic dodecan 2-yl sulfate and tetradecan 2-yl sulfate. Of course, with all racemic alkan 2-yl sulfates only the D-form was hydrolysed. Moreover, experiments with D- and L- octan 2-yl sulfates established that the L-form served as a competitive inhibitor of enzyme activity toward the D-form (K_i = 3.6 m*M*, K_m for the D-form = 1.3 m*M*). Under these circumstances it is clear that meaningful kinetic data on the enzyme cannot be obtained with racemic substrates.

For this reason a number of racemic alkan 2-ols (C_7 to C_{14}) were resolved into D- and L- forms and the sulfate esters of the former were tested as substrates for the CS2 enzyme. Because of problems that develop with the assay of enzyme activity when substrate chain length approaches C_{10}, the C_{10}, C_{12}, and C_{14} esters were labeled with ^{35}S and a new enzyme assay method was developed for these particular compounds. Using these defined D-enantiomorphs it was shown that enzyme/substrate affinity increased with increasing chain length, and a plot of log K_m against the number of carbon atoms in the alkyl chain gave the straight line relationship characteristic of an homologous series.[35] Values of V_{max} increased as chain length increased from C_7 to C_9, and thereafter rapidly decreased with increasing chain length (see Table 1).

In parallel experiments it was also shown that members of homologous series of L-alkan 2-yl sulfates, primary alkyl sulfates and alkanesulfonates served as competitive inhibitors of the CS2 enzyme (Table 2). Moreover, for each series of compounds, log K_i values decreased linearly with increasing alkyl chain length. The implications of the collective results in terms of the contributions of alkyl chain and ester sulfate group to the binding of substrates or competitive inhibitors are discussed at greater length in Volume II, Chapter 2. It suffices to say here that hydrophobic interactions make a major contribution to the overall binding process.

Further findings of considerable importance emerged as a result of these various studies on the CS2 alkylsulfatase. Thus, it became clear that the products of the enzyme reaction were inorganic sulfate and parent secondary alcohol as expected, but hydrolysis was accompanied by an inversion of configuration of the latter. Thus, when D-octan 2-yl sulfate served as substrate, L-octan 2-ol was liberated.[34] Other studies (unpublished) showed the absence of any racemase activity. This suggested that the enzyme was an exception to the general rule that hydrolytic enzymes acting on ester linkages cleave the acid–oxygen bond rather than the alcohol–oxygen bond. This was confirmed by allowing the CS2 enzyme to hydrolyse substrate in the presence of $H_2{}^{18}O$ and subsequently finding isotopic oxygen in the appropriate quantity in the liberated alcohol. No isotope was found in the liberated inorganic sulfate (Equation 1).

$$R{-}\!\!\mid\!\!{-}O{-}SO_3^- + H_2{}^{18}O \rightarrow R{-}{}^{18}OH + HO \cdot SO_3^- \qquad (1)$$

Hitherto, it had been tacitly assumed, following the work of Spencer,[36,37] that enzymic hydrolysis of sulfate esters proceeded via cleavage of the O–S bond of the C–O–S linkage. However, this assumption is no longer tenable and, indeed, as will emerge from later sections, all the simple alkylsulfatases so far examined operate by cleavage of the C–O bond. This finding adds a new dimension not only to the sulfatase field, but also that of ester hydrolases in general. Further comments on the mechanism of action of the CS2 and other simple alkylsulfatases are provided in Volume II, Chapter 2.

The assay substrate recommended currently for assaying the CS2 enzyme is D-octan 2-yl sulfate. Greater activity can be achieved using the corresponding D-nonan 2-yl sulfate, but the parent alcohol (unlike D-octan 2-ol) is not commercially available and must be obtained by resolution of the racemate.

Table 1
K_m AND V_{max} AVAVALUES FOR SOME SUBSTRATES OF THE CS2 ALKYLSULFATASE OF *COMAMONAS TERRIGENA*

Substrate	K_m (m*M*)	Relative-V_{max}[a]
D-Heptan-2-yl sulfate	6.80	18
D-Octan-2-yl sulfate	1.02	80
D-Nonan-2-yl sulfate	0.73	100
D-Decan-2-yl sulfate	0.25	22
D-Dodecan-2-yl sulfate	0.055	1.2
D-Tetradecan-2-yl sulfate	0.006	0.02

[a] V_{max} is expressed relative to that for D-nonan-2-yl sulfate which is taken as 100.

Table 2
COMPETITIVE INHIBITION OF THE CS2 AND S1 STEREOSPECIFIC SECONDARY ALKYLSULFATASES BY VARIOUS ALKYL SULFATES AND ALKANESULFONATES

	Average K_i (m*M*)	
Inhibitor	CS2 Enzyme (*Comamonas terrigena*)	S1 Enzyme (*Pseudomonas* C12B)
Secondary alkyl sulfates		
L-Octan-2-yl sulfate	2.39	1.85
L-Nonan-2-yl sulfate	1.36	—
L-Decan-2-yl sulfate	0.30	—
L-Tetradecan-2-yl sulfate	0.02	—
Primary alkyl sulfates		
Propyl sulfate	59.0	—
Butyl sulfate	46.8	23.0
Pentyl sulfate	14.3	14.0
Hexyl sulfate	6.4	6.0
Heptyl sulfate	3.1	1.9
Octyl sulfate	1.3	0.9
Nonyl sulfate	0.73	0.4
Decyl sulfate	0.30	0.14
Alkanesulfonates		
Butanesulfonate	161.5	—
Pentanesulfonate	66.0	—
Hexanesulfonate	19.2	29.0
Heptanesulfonate	8.9	7.9
Octanesulfonate	4.5	2.4
Nonanesulfonate	1.6	0.6

Note: D-Octan-2-yl sulfate was employed as substrate.

Finally, attention should be drawn to the fact that the CS2 enzyme is localized in or near the peptidoglycan region of the bacterial cell[32] and further comment on this point is made in Volume II, Chapter 4. Some general properties of the enzyme (including standard assay conditions) are summarized and compared with those of other simple alkylsulfatases in Table 3.

Table 3

SOME PROPERTIES OF THE ALKYLSULFATASES OF *COMAMONAS TERRIGENA* AND *PSEUDOMONAS* C12B

	C. terrigena enzymes		*Pseudomonas* C12B enzymes				
	CS1	CS2	S1	S2	S3	P1	P2
Approximate molecular weight	>250,000	250,000	>250,000	—	46,000	342,000	160,000
Molecular weight by SDS/ polyacrylamide gel	—	57,000	—	—	13,000	88,000	79,400
Recommended assay substrate	L-octan-2-yl sulfate	D-octan-2-yl sulfate	D-octan-2-yl sulfate	L-octan-2-yl sulfate	Nonan-5-yl sulfate	Octyl sulfate	Octyl sulfate
Optimum concentration of assay substrate	15 m*M*	15 m*M*	7m*M*	15 m*M*[a]	15 m*M*	25 m*M*	3 m*M*
Optimum pH	7.3	7.5	7.1	7.1[a]	7.5	7.0	8.3
Buffer	0.1 *M*-Tris/ maleate	0.1 *M*-Tris/ maleate	0.1 *M*-Tris/ HCl	0.1 *M*-Tris/ HCl	0.1 *M*-Tris/ HCl	0.05 *M*-Tris/ HCl	0.1 *M*-Tris/ HCl
Cellular localization	Peptidoglycan	Peptidoglycan	Periplasmic	Periplasmic	Cell periphery	Periplasmic	Periplasmic
C–O bond cleavage	Probably	Yes	Yes	Probably	Yes	Probably	Yes
Inversion of configuration	Yes	Yes	Yes	Probably	Yes[b]	Not applicable	Not applicable

[a] Conditions employed but not necessarily optimal.

[b] When acting on L-alkan-2-yl sulfates.

C. The CS1 Secondary Alkylsulfatase of *Comamonas terrigena*

At the time of writing no serious attempt has been made to purify this enzyme to homogeneity. During the purification of the coexisting CS2 enzyme (see Section II. B. 1.) the CS1 enzyme exhibited a considerable degree of instability. However, it proved possible to separate the remaining activity from that of the CS2 enzyme at the final stage of purification of the latter. CS1 was recovered in 1% yield and had been purified 67-fold. No determinations of molecular weight have been made, but the enzyme eluted from Sephadex® G-100 during the purification procedure in a band that overlapped but was slightly ahead of that due to the CS2 enzyme.[32] This suggests that the molecular weight of the CS1 enzyme is greater than 250,000. No very detailed studies on this preparation have been made, but it is known that the enzyme, in direct contrast to the CS2 enzyme, is stereospecific for L-alkan 2-yl sulfates.

Maximum activity toward 15 m*M* L-octan 2-yl sulfate in 0.1 *M* Tris-maleate buffer was obtained at pH 7.3, and under these buffer and pH conditions the K_m for the enzyme acting on this substrate was 2.8 m*M*. D-Octan 2-yl sulfate was a competitive inhibitor of the enzyme with a K_i of 5.5 m*M*.[38] Inversion of configuration occurred during the hydrolysis of L-octan 2-yl sulfate by the enzyme and it is assumed that a C−O bond split is again involved. However, no attempt has yet been made to confirm this by experiments with $H_2{}^{18}O$. The enzyme is also localized in or near the peptidoglycan region of the bacterial cell.[32]

When the CS1 and CS2 enzymes are both present as, for example, in cell extracts of *C. terrigena*, a substrate such as racemic octan 2-yl sulfate is completely degraded in spite of the fact that each of the enzymes is inhibited by the substrates of the other. However, the difference in each case between the affinity of enzyme for substrate and that for inhibitor is such that hydrolysis can proceed.

D. The S1 Secondary Alkylsulfatase of *Pseudomonas* C12B

In dealing with this enzyme it is necessary to remind the reader (see Section II. B.) that, in terms of properties, the S1 and S2 enzymes of *Pseudomonas* C12B actually correspond, respectively, to the CS2 and CS1 enzymes of *C. terrigena*.

1. Purification and General Properties

Pseudomonas C12B is maintained on nutrient agar slants and has been so maintained since 1963 without any apparent change in characteristics and properties. For studies on the S1 enzyme the organism was grown at 30° C to the stationary phase on a large scale (350 ℓ) on nutrient broth. Freeze-dried ruptured cells retained activity of the enzyme for several months when stored at −20° C. Development of a method for purifying the enzyme to homogeneity provided a great deal of frustration, mainly because of the problems of separating the enzyme from the coexisting S2 secondary alkylsulfatase and the P1 primary alkylsulfatase. A method involving the following steps was finally developed.[39]

Nucleic acid was removed from an extract of the freeze-dried cells by treatment with streptomycin sulfate and this step was followed by treatment of the dialysed, nucleic acid-free extract wth ammonium sulfate to give 55% saturation. The protein that precipitated contained much of the P1 primary alkylsulfatase activity but little or no S1 enzyme activity, nor could the latter enzyme be induced to precipitate in significant amounts when the concentration of ammonium sulfate was greatly increased. The whole was therefore dialysed and subsequently concentrated. Fractionations on DEAE-cellulose and Sephadex® G-100 followed and, finally, a much closer fractionation on the former was repeated and yielded pure enzyme. The purification achieved was approximately 5,000-fold with retention of about 5% of the original activity. The S2 secondary alkylsulfatase showed a degree of instability somewhat similar to that shown by the analogous CS1 enzyme from *C. terrigena*. Remaining traces of the S2 enzyme were eliminated during the final purification stage.

The molecular weight of the enzyme as determined by gel filtration was 285,000, but there is no information available concerning the quaternary make-up of the enzyme. Investigations[39,40] similar to those made on the CS1 and CS2 enzymes[32] established that the S1 enzyme was stereospecific for D-alkan 2-yl sulfates and activity toward D-octan 2-yl sulfate was competitively inhibited by the corresponding L-isomer (K_i = 1.85 m*M*, Table 2). However, the K_m for enzyme activity toward D-octan 2-yl sulfate is 0.78 m*M*, so that hydrolysis of the D-component of a racemic alkan 2-yl sulfate can still proceed.

Primary alkyl sulfate esters and alkanesulfonates also inhibited the enzyme competitively (Table 2), inhibition increasing with increasing chain length. With each series a straight line relationship (characteristic of homologous series) was obtained when log K_i was plotted against chain length, and further comments on the significance of the alkyl chain in the binding of inhibitors and substrates are presented in Volume II, Chapter 2. The general similarity between this S1 enzyme of *Pseudomonas* C12B and the CS2 enzyme of *C. terrigena* is further emphasized by the finding[34] that the former enzyme also catalyzes the rupture of the C—O bond of the C—O—S linkage. Inversion of configuration accompanies this, so that the liberated alcohol has the L-configuration.

Because of the commercial availability of D-octan 2-ol, D-octan 2-yl sulfate is the recommended substrate for the S1 enzyme and activity toward this substrate at a concentration of 7 m*M* is measured at 30° C in the presence of 0.1 *M* Tris/HCl buffer, pH 7.1 (see Table 3). The enzyme is localized in the periplasmic region of the bacterial cell.[41]

E. The S2 Secondary Alkylsulfatase of *Pseudomonas* C12B

At the time of writing no serious attempt to purify this enzyme has been made. However, the enzyme has been separated from the S1 enzyme by polyacrylamide gel electrophoresis and the appropriate band of activity recovered in low yield.[39] This preparation rapidly hydrolysed L-octan 2-yl sulfate completely, but only feeble activity was exhibited toward the corresponding D-isomer and this was considered to be due to a trace of contaminating S1 enzyme. The enzyme also hydrolysed racemic decan 2-yl sulfate to the extent of 50%, but no activity was observed toward heptan 4-yl sulfate or nonan 5-yl sulfate.

Clearly much more work needs to be done on the S2 enzyme, but it seems clear that it is analogous to the CS1 in being stereospecific for D-alkan 2-yl sulfates. The enzyme shares a periplasmic localization with the S1 enzyme.[41]

F. The S3 Secondary Alkylsulfatase of *Pseudomonas* C12B

This is perhaps the most fascinating of the three secondary alkylsulfatases of *Pseudomonas* C12B. It will be recalled from the introduction to this chapter that the enzyme is induced when the microorganism is grown on nutrient broth containing the secondary alkyl sulfate detergent Oronite®. However, it became clear[29] that the detergent formulation contained both free and sulfated long-chain secondary alcohols and it eventually emerged that induction depended on the presence in the growth medium of any combination of ester and alcohol from "tetradecan 2-yl sulfate", "hexadecan 2-yl sulfate", tetradecan 2-ol, and hexadecan 2-ol. Further comment on this point is made in Volume II, Chapter 3, but it is worth noting here that this discovery apparently represented the first report of an enzyme system that required a combination of two inducers for its production. Other systems requiring more than one inducer have since been described.[42,43]

1. Purification and General Properties

Pseudomonas C12B was grown to the stationary phase on a large scale (350 ℓ) on nutrient broth supplemented with 1.0 m*M* "tetradecan 2-yl sulfate" and tetradecan 2-ol. Freeze-dried, ruptured cells could be stored at −20° C without loss in activity. The S3 enzyme was purified from cell extracts by a procedure[44] that involved removal of nucleic acid with streptomycin sulfate, fractionations on DEAE-cellulose, Sephadex® G-100 and Blue Sepharose® CL-6B, and refractionation on Sephadex® G-100. The enzyme was purified about 680-fold with recovery of 21% of the initial activity. The final preparation was virtually homogeneous as evidenced by polyacrylamide gel electrophoresis, but a second faint band could be detected if high concentrations of protein were loaded on to gels.

The molecular weight of the enzyme was 40,000 as measured by Sephadex® G-100 gel filtration and 46,000 by electrophoresis on various concentrations of polyacrylamide gel under nondenaturing conditions. SDS-polyacrylamide gel electrophoresis gave an intense band corresponding to a molecular weight of 13,000 with two additional (but very faint) bands with molecular weights corresponding to 17,000 and 49,000. These results suggest a quaternary structure for the enzyme, but the precise meaning of the results is not clear.

Enzyme specificity studies[40,44] showed that unlike the coexisting S1 or S2 enzymes, the S3 alkylsulfatase was active toward symmetrical substrates such as pentan 3-yl sulfate, heptan 4-yl sulfate, and nonan 5-yl sulfate. The affinity of enzyme for these substrates increased with increasing chain length of the latter. Nonan 5-yl sulfate is preferred for routine enzyme assay (see Table 3 for assay conditions).

A number of asymmetric secondary alkyl sulfates were also tested as possible substrates for the S3 enzyme. Reaction progress curves revealed that a preparation of racemic octan 4-yl sulfate was readily hydrolysed to completion, showing that both D- and L-forms of the compound were substrates for the enzyme. However with racemic octan 3-yl sulfate, 50% of the ester was rapidly hydrolysed but the remainder only at a very much reduced rate. Addition of more enzyme to incubation mixtures immediately following the rapid hydrolysis phase merely accelerated the slow phase slightly, but addition of more substrate resulted in rapid hydrolysis of 50% of the amount added. It seemed logical to assume that one of the substrate stereoisomers was a good substrate for the S3 enzyme, whereas the other was a relatively poor one. Support for this assumption came from the finding that racemic nonan 4-yl sulfate and racemic nonan 3-yl sulfate were similarly hydrolysed by the enzyme in a biphasic manner, the rate of the second phase being almost three orders of magnitude slower than the first. None of these sulfate esters was available in resolved form. However, resolved isomers of octan 2-yl sulfate were available and the L-isomer was completely hydrolysed, although at a relatively slow rate. In contrast, hydrolysis of the D-isomer did not occur, even after prolonged incubation with enzyme. By analogy, it seems reasonable to conclude that L-forms of the other asymmetric esters tested are the forms that are being rapidly hydrolysed. As the position of the sulfate group is moved nearer to the center of the alkyl chain, the more the D-isomer is able to serve as a substrate. The implications of these findings in terms of the mechanism of action of the enzyme are discussed in greater detail in Volume II, Chapter 2.

Other interesting differences between the S3 enzyme and the other two secondary alkylsulfatases of *Pseudomonas* C12B are to be found in the behavior toward inhibitors. Using nonan 5-yl sulfate as assay substrate a series of primary alkyl sulfates (C_2 to C_{10}) were tested as potential inhibitors. Only butyl sulfate served as a competitive inhibitor, the C_2, C_3, C_5, C_6, and C_7 compounds having little or no inhibitory effects. The higher members of the series (C_8 to C_{10}) were inhibitors, but of the noncompetitive type. These findings contrast sharply with those obtained for the S1 enzyme (and the

CS2 enzyme of *C. terrigena)* which, it will be recalled, was competitively inhibited by primary alkyl sulfates with K_i decreasing with increasing chain length in a manner characteristic of an homologous series.

The question now arose as to the position of bond cleavage by the enzyme and the possible inversion of configuration of the liberated alcohol. Clearly, symmetrical substrates such as nonan 5-yl sulfate could be used to study the former but not the latter. Because of this L-octan 2-yl sulfate was used as substrate, in spite of the relatively low rate of hydrolysis. Inversion of configuration, yielding D-octan 2-ol, was confirmed by polarimetry and C–O (rather than O–S) bond splitting was demonstrated with the aid of $H_2{}^{18}O$.

The S3 alkylsulfatase is probably localized in the periphery of the bacterial cell.[45]

G. The P2 Primary Alkylsulfatase of *Pseudomonas* C12B

It will be recalled from the introduction to this chapter that *Pseudomonas* C12B is able to produce two primary alkylsulfatases that have been designated as P1 and P2 on the basis of movement toward the anode on gel electrophoretograms (see Figure 5). P1 is probably constitutive while P2 is produced in response to the presence of primary alkyl sulfates or alkanesulfonates in the growth medium.[29,46,47] At the time of writing relatively little work has been done on the P1 enzyme, but a study in some depth has been made on P2.[47,48] For this reason, the latter enzyme will be described before the former.

1. Purification and General Properties

Work from Fitzgerald's laboratory[30,41] established that the P2 enzyme had a relatively transient existence in cells grown in the presence of hexyl sulfate on either nutrient broth or on a basal salts-citrate medium. Maximum enzyme activity in the cells was obtained in the midexponential growth phase in batch cultures and thereafter rapidly declined to relatively low levels that persisted well into the stationary phase. Interestingly, when citrate (in concentrations of 1.0% or greater) was used as carbon source for growth the cellular levels of the S1 and S2 secondary alkylsulfatases (but not the primary alkylsulfatases) were markedly reduced and leakage of active enzymes into the medium could not account for more than a small portion of the loss. Fitzgerald and Laslie[41] suggested that the attachment of these enzymes to the cell wall was facilitated by magnesium and that high concentrations of citrate were capable of chelating the Mg^{2+} ions that were so involved. It was further suggested that the release of the enzymes from their cell wall association under these conditions perhaps allowed appreciable denaturation of the enzyme protein to occur. Irrespective of the reason, this effect seemed to be useful from the point of view of purifying the P2 enzyme as it would perhaps minimize the problem of separating it from the secondary alkylsulfatases. Furthermore, the maximal appearance of the P2 enzyme during the midexponential growth phase was helpful in that the P1 enzyme is present in relatively low amounts at this stage.

With these points in mind, enzyme purification studies were undertaken on cells of *Pseudomonas* C12B that had been grown on a large scale as follows.[48] The organism was grown at 30° C in 350 ℓ of a basal salts medium[28] supplemented with 1% sodium citrate and 4 m*M* potassium hexyl sulfate. Cells were harvested at midexponential growth phase (A_{420} = 0.6 to 0.7),[30] with the aid of a Sharples centrifuge capable of dealing with about 100 ℓ/hr. Wet cells were subsequently resuspended in 5 m*M* Tris/HCl buffer, pH 7.5, and ruptured with the aid of a Manton-Gaulin homogenizer. The broken cell suspension was then freeze-dried and could be stored for several months at −20° C without significant loss of P2 enzyme activity.

Two problems were inherent in this large scale procedure. First, because of the tran-

Table 4
K_m AND V_{max} VALUES FOR THE P2 PRIMARY ALKYLSULFATASE OF *PSEUDOMONAS* C12B

Substrate	K_m (m*M*)	V_{max} (μmol SO_4^{2-} liberated/min/mg protein)
Pentyl sulfate	30.3	3.6
Hexyl sulfate	12.5	9.1
Heptyl sulfate	3.77	16.3
Octyl sulfate	1.50	20.8
Nonyl sulfate	0.49	40.8
Dodecyl sulfate	0.03	7.2

sient existence of the enzyme and the need to harvest at midexponential growth phase, the yield of cells was rather less than desirable and, second, because of the time needed to harvest cells from such large volumes, the actual period of growth varied somewhat between those cells passing to the centrifuge first and those passing later. The fact that trial purification experiments with cells grown on a smaller scale (10 ℓ) could not always be repeated exactly when similar quantities of cells were used from a batch grown on a large scale, probably reflects the problems outlined above. Similarly, some variation was noted between the behavior of different large scale batches of cells and it was necessary to make minor changes in the experimental procedure (particularly, in the fractions pooled from column separations) when starting with a new batch of cells.

With this point in mind a purification procedure has been described[48] that involves treatment of a cell-free extract with streptomycin sulfate in order to remove nucleic acid, fractionation on DEAE-cellulose (DE-52) followed by fractionation on Sephadex® G-100 and hydrophobic chromatography on a butyl-agarose column. Fractions from the hydrophobic column that contained the P2 enzyme as the only protein (single band on gel electrophoretograms) were pooled and used for molecular weight and enzyme specificity determinations. Other fractions that contained the enzyme and trace amounts of a second protein were pooled and used in experiments in which absolute purity was not essential. The enzyme purification achieved varied between 200- and 250-fold, depending on the particular batch of cells used as starting material. Recoveries usually exceeded 50%, and final preparations showed no trace of the P1, S1, and S2 alkylsulfatases, as judged by the polyacrylamide-gel electrophoresis method of Payne et al.[27]

The molecular weight of the enzyme was 162,000 as determined by gel filtration and 158,000 as determined by the gel electrophoresis method of Zwann.[49] Interestingly, considerable difficulties were experienced in subjecting the enzyme to sodium dodecyl sulfate (SDS)/polyacrylamide-gel electrophoresis because the enzyme continued to use the dodecyl sulfate as a substrate during and after the usual initial mercaptoethanol treatment. This led to considerable interference with the electrophoretic movement of the enzyme protein, possibly because of the presence of enzymically liberated dodecanol. Satisfactory electrophoresis runs were obtained however, when the enzyme preparation was heated in a boiling water bath for a few seconds before the treatment with mercaptoethanol. A molecular weight of 79,400 was obtained under these circumstances, pointing to a dimeric structure for the native enzyme.

The optimum pH of the enzyme was about 8.3, irrespective of the chain length of the substrate used. All the primary alkyl sulfate esters tested (varying in chain length from C_5 to C_{12}, see Table 4) served as substrates for the enzyme with maximum activity being obtained with nonyl sulfate. All substrates gave normal Michaelis-Menten curves, except for dodecyl sulfate where there was a marked inhibition by excess substrate that appeared to reflect the formation of detergent micelles. The K_m values for the different substrates decreased as the chain length of the substrate increased. When log K_m was plotted against the number of carbon atoms in the alkyl chains, a straight line relationship was obtained, behavior characteristic of an homologous series. The enzyme was also active toward tetradecyl sulfate, as evidenced by the gel zymogram method of Payne et al.,[27] but no attempt has yet been made to assay the enzyme with this substrate. For various reasons, octyl sulfate was finally selected as the routine assay substrate (see Table 3 for assay conditions). Using this substrate it was shown that alkanesulfonates and secondary alkan 2-yl sulfates inhibited the enzyme in a competitive manner with K_i values decreasing with increasing chain length within each series in a manner characteristic of homologous series. Collectively, these results clearly indicate the importance of the alkyl chain and its relative hydrophobicity in enzyme-substrate or enzyme-inhibitor interactions. This point is developed further in Volume II, Chapter 2.

Perhaps the most surprising thing about the P2 primary alkylsulfatase emerged from a study of the nature of the bond that was ruptured as a result of the action of the enzyme. In common with the secondary alkylsulfatases that have so far been examined, it was shown that the P2 primary enzyme catalyzed the fission of the C—O bond of the C—O—S linkage.[50] As will be seen later (Volume I, Chapter 4). *Pseudomonas* C12B has the capacity to produce an arylsulfatase enzyme under certain growth conditions. Using *p*-nitrophenyl sulfate as substrate in the presence of $H_2{}^{18}O$, it was possible to demonstrate the appearance of the ^{18}O-label in the sulfate liberated as a result of arylsulfatase action. Thus, in contrast to the alkylsulfatase enzymes, but in common with other arylsulfatase enzymes that have been so examined, the *Pseudomonas* C12B enzyme catalyzes the rupture of the O—S bond of the C—O—S linkage. Further comment on these findings is made in Volume II, Chapter 2.

Some preliminary attempts have been made to examine the active site of the P2 alkylsulfatase enzyme, but the only observation of any significance has been the demonstration that basic residues (guanidinum, imidazole, or lysine ε-amino groups) do not appear to be accessible for modification at the active site.

In common with the S1, S2, and P1 alkylsulfatases of *Pseudomonas* C12B, the P2 enzyme is localized in the periplasmic region of the bacterial cell.[41]

Finally, the question might be posed as to why hexyl sulfate was used as the P2 enzyme inducer when the bacterium was grown on the large scale. As described in Volume II, Chapter 3, the induction of this enzyme in relation to the nature, chain length, and concentration of the inducer is somewhat complex. Experiments with resting cells show that primary alkyl sulfates are generally more efficient inducers for the synthesis of the enzyme than alkanesulfonates of similar chain length.[47] However, higher concentrations of the former compounds are required and this presumably reflects the fact that the alkyl sulfates are nongratuitous inducers that are substrates for both the P1 and P2 enzymes. For example, resting cells (separated at midexponential growth phase) shaken with 4 m*M* octyl sulfate for a period of 5 hr, not only produced the P2 enzyme, but also hydrolysed almost 70% of the inducer. On the basis of resting cell experiments in which alkyl sulfates were tested as inducers at varying concentrations, a comparison of the effect of each alkyl sulfate at its optimal inducer concentration showed that maximum induction was obtained with 4 m*M* dodecyl sulfate (Figure 6). However, attempts to grow cells on the basal salts-sodium citrate medium containing dodecyl sulfate at that concentration were unsatisfactory because of the extremely slow growth rate obtained. A series of other alkyl sulfates were similarly tested in growth experiments (unpublished results, Cardiff laboratories), but hexyl sulfate gave the most satisfactory results. This was an interesting result in that Fitzgerald and Payne[28] had originally selected this ester as the most suitable inducer on a somewhat arbitrary basis.

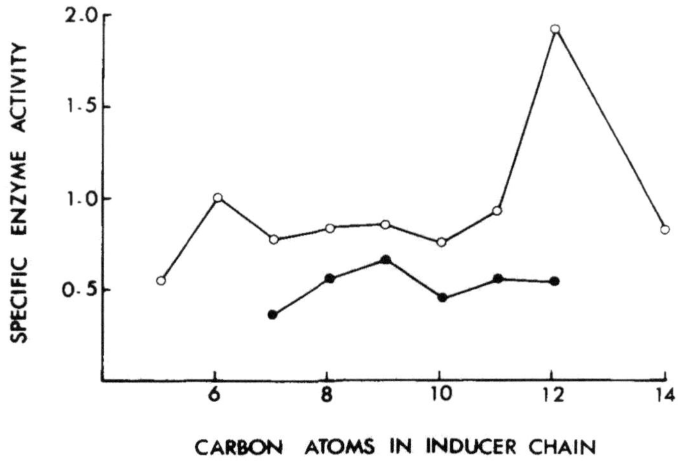

FIGURE 6. Relationships between carbon chain length of inducer and synthesis of the P2 primary alkylsulfatase of *Pseudomonas* C12B after exposure of resting cells to optimum concentration of each inducer. ○, Primary alkyl sulfates; ●, alkanesulfonates.

H. The P1 Primary Alkylsulfatase of *Pseudomonas* C12B

Very little work on this enzyme has been done at the time of writing and the following short account provides details of the current situation.

Experiments with crude extracts of cells grown on nutrient broth have established that the enzyme has very little detectable activity toward butyl sulfate, but shows increasing activity as the chain length of the substrate increases from pentyl sulfate through to nonyl sulfate. The crude enzyme also hydrolyses dodecyl sulfate as evidenced from the fact that this ester is normally used as the substrate for detecting the enzyme on gel zymograms. The immediate impression, therefore, is that the P1 enzyme exhibits the same specificity as the P2 enzyme. Octyl sulfate has been employed as assay substrate during preliminary purification studies and the enzyme exhibits maximum activity toward 25 m*M* octyl sulfate at pH 7.0 in the presence of 0.05 *M* Tris/HCl buffer.

First attempts to obtain a pure enzyme have involved treatment of a cell-free extract of broth grown cells with streptomycin sulfate, followed by fractionation on DEAE-cellulose (DE-52), and two fractionations on Sephadex® G-100 Superfine. The purification achieved was greater than 3000-fold with a recovery of 25% of the starting activity. However, the preparation still contained small amounts of the S1 and S2 secondary alkylsulfatases (about 0.3 % of the amounts originally present) as well as trace amounts of at least one other protein.

This preparation was used for the purpose of molecular weight determinations. Gel-filtration on Sephadex® G-200 gave a molecular weight of 338,000; polyacrylamide gel electrophoresis used as described by Zwann[49] gave a molecular weight of 346,000, and SDS/polyacrylamide gel electrophoresis (after initial treatment of the enzyme at 100° C for a few seconds; see previous section) gave a molecular weight of 88,000. It therefore appears probable that the enzyme is a tetramer, thereby distinguishing it from the P2 enzyme which, it will be recalled, has a much lower molecular weight (approximately 160,000) and appears to be a dimer.

At the present time it is not absolutely clear whether the P1 enzyme is truly a constitutive enzyme. The enzyme is always present in broth grown cells, but this does not exclude the possibility that something present in the broth is serving as an inducer.

Moreover, in some of the early experiments on the *Pseudomonas* C12B alkylsulfatases[46] there were indications that some growth conditions that excluded broth gave cells that did not possess the P1 enzyme. For the present the authors continue to regard P1 as constitutive, but further study on this point clearly needs to be made.

I. General Comments on the *Pseudomonas* C12B Alkylsulfatases

Although much remains to be done, it is already clear that *Pseudomonas* C12B is beautifully adapted (cf. Table 3) to deal with the initial step in the biodegradation of primary alkyl sulfate surfactants such as sodium dodecyl sulfate, as well as detergent preparations such as Oronite® that are compounded of a variety of secondary alkyl sulfates in which the position of the sulfate ester group is not necessarily at C-2. The method of preparation of secondary alkyl sulfates by, for example, the treatment of commercial alkene mixtures with sulfuric acid, or the treatment of secondary alcohols with sulfuric acid, generally leads to a predominance of racemates of C-2 sulfate esters, irrespective of the positions of the unsaturated linkages (or hydroxyl groups) in the starting materials.[33] However, the secondary alkylsulfatase complement of *Pseudomonas* C12B under circumstances where a detergent such as Oronite® is present, ensures that not only the predominant D- and L-alkan 2-yl-sulfate components, but also most of the minor stereoisomeric or symmetrical components in which the sulfate group occupies a position other than C-2, could be effectively desulfated. Whether or not any components of such a detergent remain relatively resistant to attack by the secondary alkylsulfatase complex has still not been established. However, a study of the results of tests of detergent biodegradation such as the river water die-away test,[51] do suggest that not all the components of a secondary alkyl sulfate detergent are biodegraded at the same rate. Moreover, it will be recalled that the D-forms of secondary alkyl sulfates in which the sulfate group was at C-3 were hydrolysed extremely slowly. It will now be interesting to examine these problems further at the level of enzyme specificity and to follow the initial degradation step of a detergent such as Oronite® in order to identify components that are relatively rapidly hydrolysed or poorly hydrolysed by cell-free extracts of *Pseudomonas* C12B that contain the full complement of secondary alkylsulfatases.

The ability of *Pseudomonas* C12B to produce two primary alkylsulfatases is interesting but puzzling. At the time of writing, there is some evidence to suggest that the specificities of the two enzymes are similar if not identical. Is, therefore, the presence of the two enzymes in the bacterium purely fortuitous? For example, is one of them plasmid coded and the other chromosomal? Whatever the reason for the copresence of the enzymes, it would seem to be of the utmost importance to subject both enzymes to detailed specificity studies. It was pointed out in Chapter 1 that some individual sulfatase enzymes (e.g., arylsulfatase A of mammalian lysosomes) show activity toward very different types of sulfate ester. It was further pointed out at the beginning of the present chapter, that several naturally occurring compounds could possibly serve as substrates for primary alkylsulfatases. It is therefore possible that one or other of these enzymes alone shows specificity toward one or more of these natural substrates. Herein lies a most fascinating area for future study.

A further interesting point concerns the relationship between the five enzymes and several points immediately come to mind. There is clear evidence with the S1, S3, and P2 enzymes (and preliminary indications with P1) that hydrophobic interactions are important to the binding of substrate to enzyme. Do similar hydrophobic clefts therefore exist in each of the enzymes? Some support for this possibility comes from the fact that substrates for the P2 enzyme are competitive inhibitors of the S1 enzyme and vice versa. The modes of action of the S1, S3, and P2 enzymes are similar in terms of position of bond cleavage and (in the cases of S1 and S3) inversion of configuration;

and it is highly probable that similar considerations will eventually be shown to apply to the S2 and P1 enzymes. Do the catalytic sites of the enzymes therefore exhibit similar features? Interestingly, the molecular weights that have been determined so far (Table 3) suggest considerable differences between the enzymes at the macromolecular level. P1 emerges as the largest with a molecular weight approaching 350,000. It is difficult to envisage a macromolecular relationship between this enzyme and the S3 enzyme (molecular weight of the order of 45,000). On the other hand a relationship between the tetrameric P1 and the dimeric P2 primary alkylsulfatases seems less unreasonable, particularly as the molecular weight of the former is approximately twice that of the latter.

Of course, the genetic relationship, if any, between the five enzymes also represents a subject of considerable intrinsic interest. The coexistence of such a number of hydrolytic enzymes with activity toward different isomers of one class of compound provides us with a unique situation in the field of molecular enzymology that must surely be the subject of a future research program.

J. Other Microbial Sources of Simple Alkylsulfatases

It seems likely that alkylsulfatases are already present or can be induced in many different bacteria apart from the two (*Pseudomonas* C12B and *C. terrigena*) that have been studied in detail at the enzymological level. The presence of an inducible primary alkylsulfatase in strains of *P. aeruginosa* has been noticed by several workers,[17,52-57] but apart from the enzyme induction studies described in Volume II, Chapter 3, no detailed work on the enzyme has been reported. Workers in the U.S.S.R. have made some useful studies[53,54] on the distribution of the ability of bacteria present in collections to use primary alkyl sulfates as sole source of carbon and energy and have shown that many *Pseudomonas* species possess this ability (see also Hsu[19,22]). Of 191 strains tested, representing 23 species of the genus, all strains of *P. aeruginosa, P. aureofaciens,* and most strains of *P. alcaligenes, P. pseudoalcaligenes, P. mendocina,* and *P. aurantiaca,* were active in this respect.[54] Activity was also detected in representatives of other species of bacteria, including *Achromobacter guttatas, Flavobacterium devorans, Citrobacter freundi, Aerobacter* (now *Enterobacter) aerogenes, Nocardium corallina,* and *Aerobacter cloacae.*[53] The periplasmic localization of primary alkylsulfatase activity in one of the pseudomonads examined (*Pseudomonas* 2T/L)[58] was similar to that of the P2 enzyme of *Pseudomonas* C12B.[41] Other workers[52,59] have detected primary alkylsulfatase activity in *Klebsiella* species, including *Klebsiella aerogenes.*

Searches for organisms with secondary alkylsulfatases have been more fragmentary. Apart from the presence of such enzymes in *Comamonas terrigena* and *Pseudomonas* C12B, activity toward secondary alkyl sulfates has been detected in *A. cloacae,*[26] *Pseudomonas* B-2 (presumed but not stated directly[60]), and as a component present in mixed cultures of microorganisms from activated sludge.[61] In the last case it is of some interest that the branched primary alkyl sulfates, 2-ethylhexyl sulfate and 2-methylundecyl sulfate, were also both hydrolysed by extracts of cells of the mixed flora that had been grown on sodium dodecyl sulfate as sole source of carbon and energy. This appears to be the first report of the hydrolysis of a branched chain primary alkyl sulfate.

In almost all these cases no attempt seems to have been made to examine whether broth grown cells contain primary or secondary alkylsulfatases, or to establish the number of alkylsulfatases present in cells grown on alkyl sulfates. The results of the specificity studies with *Pseudomonas* C12B, together with the availability of chemically defined substrates and the gel zymogram method, should now allow a more methodological approach to surveys of alkylsulfatase distribution in microorganisms. Indeed, a further and as yet unidentified microorganism has been isolated in Cardiff which, when grown on nutrient broth in the absence of inducer, forms three primary alkylsulfatases and three secondary alkylsulfatases, in addition to an enzyme that removes the sulfate group from 2-(2,4-dichlorophenoxy)ethyl sulfate (see Section IV). These preliminary results suggest that perhaps the full story of the bacterial alkylsulfatases may be even more complicated than it already appears to be. Some further point is given to this possibility by the finding that the caecal contents of conventional rats were able to liberate inorganic sulfate from deoxycorticosterone 21-sulfate and certain other sulfated steroids, whereas caecal contents of germ-free rats did not possess this ability.[62] Deoxycorticosterone 21-sulfate contains a primary alkyl sulfate grouping and the implication is that some bacterial alkylsulfatase enzyme is responsible for this activity. Whether this enzyme is similar to one or other of the simple alkylsulfatases described in earlier sections of this chapter must now await the outcome of future investigations.

K. Ultimate Fate of the Liberated Alcohols

Brief mention has already been made (Section II.A) of the existence of bacterial alcohol dehydrogenase enzymes that continue the biodegradation of the alcohols that are liberated as a result of alkylsulfatase action. Presumably a primary alcohol such as dodecanol is converted via the action of the alcohol dehydrogenase to the corresponding fatty acid which can then be degraded in the usual manner. Both *C. terrigena*[63] and *Pseudomonas* C12B[23] contain NAD-dependent dehydrogenases active toward primary alcohols. The pathway of degradation of secondary alcohols liberated as a result of secondary alkylsulfatase action is more obscure. However, on the basis of studies on a pseudomonad isolated by enrichment culture in the presence of butan 2-ol, it has been proposed[60] that the degradation of an alcohol such as decan 4-ol proceeds via the ketone, hydroxyketone, and dione to a mixture of alkyl aldehyde and fatty acid.

Following the discovery of the stereospecific nature of alkylsulfatase enzymes in *C. terrigena* and *Pseudomonas* C12B, it was natural that attention should turn to the possibility of stereospecific alcohol dehydrogenases. Only the situation in *C. terrigena* has been studied at the time of writing, but the results[63] obtained so far have opened up an entirely new field of research. Briefly, the bacterium possesses two constitutive NAD-dependent alcohol dehydrogenases, one of which is active toward both primary alcohols and secondary alcohols of the D-configuration. The other is active toward secondary alcohols of the L-configuration and appears to be the first such enzyme that has been described. Specificity studies on crude cell extracts show that this enzyme is active toward L-alkan 2-ols and there is evidence to indicate a somewhat reduced activity toward symmetrical secondary alcohols such as heptan 4-ol and nonan 5-ol. In short, the presence of stereospecific secondary alkylsulfatases in *C. terrigena* cells is paralleled by the presence of stereospecific alcohol dehydrogenases. It will now be of great interest to extend these studies to *Pseudomonas* C12B with its greater complexity of alkylsulfatase complement. Attempts to purify the novel L-alcohol dehydrogenase of *C. terrigena* are now in progress and it is clear that new research horizons are in view.

III. CHOLINE SULFATASE

This enzyme has a somewhat unique place in the history of the sulfatase enzymes in that it was the first such enzyme to which a physiological role was ascribed with some degree of confidence. This role (described in greater detail in a later section) envisaged the enzyme's substrate as a fungal store of sulfur that could be hydrolysed to yield inorganic sulfate when other sources of sulfur were in short supply. Various biochemical, enzymological, and genetic studies allowed this role to be proposed, although it

I.

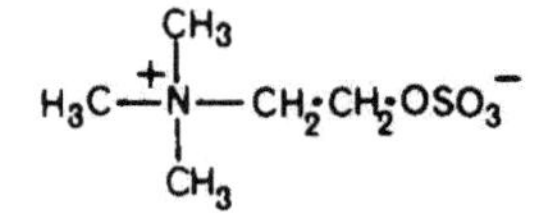

II.

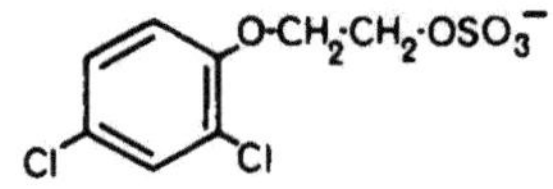

FIGURE 7. Respective substrates for choline sulfatase and phenoxyethyl sulfatase. I, choline sulfate; II, 2-(2,4-dichlorophenoxy)ethyl sulfate.

will later become apparent that other interpretations can be placed on the results of these studies. Meanwhile, it is interesting that no one has yet purified a choline sulfatase to homogeneity, a fact that appears to reflect the instability of the enzyme.

A. The Substrate, Choline Sulfate

Choline sulfate is one of the most interesting naturally occurring sulfate esters. Because it possesses very strong positively and negatively charged groups (Figure 7) positioned at a suitable distance from each other, it exists as an internally compensated salt and, as such, exhibits no movement on electrophoresis under normal conditions nor is the ester retained by mixed-bed ion-exchange resin chromatography. This property had been useful to investigators as it enables the ester to be separated readily from either inorganic sulfate or from other charged materials.

Choline sulfate was synthesized in the laboratory as long ago as 1904,[64] but more than 30 years elapsed before Woolley and Peterson[65] detected substantial amounts of the compound in the mycelia of *Aspergillus sydowi.* Sporadic reports of its occurrence elsewhere came from time to time. The "ethereal sulfate" that was detected in several species of *Aspergillus* and *Penicillium*[66] and was claimed to constitute 40% of the total mycelial sulfur in *Penicillium notatum,*[67] was almost certainly choline sulfate. The ester was subsequently isolated from *Penicillium chrysogenum*[68,69] and was also found in lichens of the genus *Rocella,*[70] in the red alga *Gelidium cartilaginum,*[71] and in higher plants.[72]

During the course of these and other investigations it slowly became apparent[67,69,73] that various fungi could grow on choline sulfate as their sulfur source and this, in turn, led to further searches for microbial sources of the ester and for enzyme systems that might be involved in its synthesis and degradation. Harada and Spencer[74] examined 9 bacterial and 32 fungal isolates for their ability to produce choline sulfate when inorganic sulfate was the sole source of sulfur in the growth medium. None of the bacteria tested were apparently able to synthesize choline sulfate and production in the fungi was confined to higher fungi. Thus it was not produced in any of five species of the Phycomycetes tested. Of the Ascomycetes, the Endomycetales (six species) were not producers, but the Sphaeriales (two species, four strains) were. A Basidiomycete and all the Fungi Imperfecti (16 species tested) except *Torulopsis utilis* produced the ester. This exercise was subsequently considerably extended by Itashi[75] who unsuccessfully searched for choline sulfate in two additional bacterial species, but supported and extended the findings of Harada and Spencer[74] with regard to fungal sources.

As a result of these investigations it seemed to be generally assumed that bacteria were unable to produce choline sulfate and the discovery[76] of a choline sulfatase in a strain of *Pseudomonas nitroreducens* did nothing to disturb this view. More than a decade was to elapse before the ester was detected as an intracellular component of *Lactobacillus plantarum*[77] and as an extracellular product of *Pseudomonas* C12B and other *Pseudomonas* species.[78]

A number of studies[78-83] have shown that the biosynthesis of choline sulfate in fungi and in bacteria involves the formation of 3′-phosphoadenosine 5′-phosphosulfate (PAPS) and the transfer of sulfate from that compound to choline via the agency of choline sulfotransferase (EC 2.8.2.6). Fungi that can utilize inorganic sulfate as a source of sulfur for growth, but lack the ability to produce choline sulfate, do not possess the necessary choline sulfotransferase.[80] Although it appears never to have been positively established, it seems to have been generally assumed that the synthesis of the ester in fungi takes place in the cytoplasm. However, it is of some interest that the sulfation of choline in *Pseudomonas* C12B appears to be mainly cell wall associated and, indeed, most of the activity appears in the culture medium during the growth of the bacterium.[83] Fungal choline sulfotransferase appears to be an enzyme of high specificity. The *A. nidulans* enzyme is specific for PAPS as sulfate donor and only choline, dimethylaminoethanol, and dimethylethylaminoethanol are known to be able to serve as sulfate acceptors.[81] Similar high specificity is shown by the enzyme present in *P. chrysogenum*,[82] but there is as yet no information on the specificity of the *Pseudomonas* C12B enzyme.

These few comments on the formation of choline sulfate have been included here because of their relevance to information presented in the following sections.

B. The Enzymes

Searches for choline sulfatases stemmed from the observations mentioned previously that various fungi could grow under conditions in which choline sulfate was serving as sole sulfur source. Clearly the sulfur could have been made available to the organisms in a number of possible ways; for example, by degradation of the organic moiety of the ester, by enzymic hydrolysis of the ester linkage, or by direct transfer from choline sulfate to an acceptor such as 3′-phosphoadenosine 5′-phosphate (PAP). In relation to the last possibility it was known that aryl sulfotransferase (EC 2.8.2.1), the enzyme responsible for the transfer of the sulfate group from PAPS to phenolic acceptors could, under certain circumstances, catalyse the transfer of sulfate from *p*-nitrophenyl sulfate to PAP. This ability was a reflection of the high group potential associated with the ester sulfate linkage of this particular aryl sulfate ester.[84] The participation of choline sulfate and choline sulfotransferase in a similar transfer of sulfate to PAP would imply a high group potential associated with the ester sulfate group of this substrate, a difficult proposition to accept in the light of its relative stability toward acid hydrolysis. In spite of this, Spencer and Harada,[80] as a result of studies on a number of fungi, strongly supported the proposition. Their view was strengthened by their failure to detect a choline sulfatase enzyme in any of the fungi and by the finding[74] that choline sulfate could act as a source of choline for a cholineless mutant of *Neurospora crassa*, thereby establishing that the carbon moiety of the ester was not degraded.

Subsequently, came the discovery of an inducible choline sulfatase in cell extracts of *P. nitroreducens*,[76] while further sources of the enzyme were found in the fungus, *Penicillium chrysogenum*,[85] and in the bacterium *Pseudomonas aeruginosa*.[86] This was followed by a convincing demonstration by Orsi and Spencer[81] of the inability of choline sulfotransferase to catalyze the transfer of sulfate from choline to PAP. Detailed studies of the mechanism of utilization of choline sulfate by various mutants of *A. nidulans* led Spencer and his colleagues to postulate that the breakdown of the ester occurred by simple hydrolysis involving a choline sulfatase.[87] The enzyme had previously been missed because of the need to derepress synthesis in the mycelial system producing the enzyme by transferring the growing organism for a period to a sulfur-free medium. Finally, choline sulfatase was also detected in *K. aerogenes,*[88] and in a pseudomonad designated as *Pseudomonas* V-A.[89]

The various fungal and bacterial enzymes appear to fall into three distinct types with respect to their synthesis:

1. The choline sulfatases of *K. aerogenes* and *Penicillium chrysogenum* appear to be constitutive enzymes.
2. The enzymes of *Pseudomonas aeruginosa, Pseudomonas* V-A, and *P. nitroreducens,* all of which are induced by the substrate, choline sulfate,
3. The derepressible enzymes of *A. nidulans* and *N. crassa.*

Further information about the control of the synthesis of these various enzymes is provided in Volume II, Chapter 3.

1. The Choline Sulfatase of Pseudomonas nitroreducens

The strain of *P. nitroreducens* in which choline sulfatase was first discovered was isolated from soil by Takebe.[76] Choline sulfatase activity was found in cells (but not growth medium) grown on a basal salts mixture containing choline sulfate as carbon, energy, and sulfur source. No enzyme activity could be detected in cells in which choline served as carbon and energy source and inorganic sulfate as sulfur source. Attempts to purify the enzyme from cell extracts were bedeviled by the instability of the enzyme and the investigator had to be content with a fourfold purification obtained by ammonium sulfate fractionation. The enzyme preparation was shown to produce equimolar amounts of choline and sulfate from its substrate and was without action toward aryl sulfates, D-glucose 6-sulfate, ethyl sulfate, the sulfated thioglycoside sinigrin, cellulose polysulfate, and chondroitin sulfate. There was a suggestion of some release of sulfate from amylose polysulfate, but the significance of this is not apparent. It was also established that the enzyme was not an acetylcholinesterase exhibiting dual specificity. The effect on enzyme activity of a variety of potential inorganic inhibitors was tested and sulfite proved to be a strong inhibitor. Some properties of the enzyme are recorded in Table 5, perhaps the most interesting one being the relatively high K_m (40 m*M*) exhibited by the enzyme.

2. The Choline Sulfatase of Pseudomonas aeruginosa

Preliminary experiments with this organism established that appreciable variation in the choline sulfatase levels could be obtained, depending on whether the bacterium was presented with choline sulfate as a carbon, nitrogen, and sulfur source or as a source for only one or two of the elements.[90] Activity was greatest in cells utilizing the ester for carbon or some combination of carbon plus nitrogen or sulfur. It was lower when the ester was used as a nitrogen source and lowest (only about one sixth of the activity of that when utilized for C, N, and S) when used as sulfur source. No enzyme appeared when cells were grown on butyl sulfate, *p*-nitrophenyl sulfate, choline phosphate, or on choline plus inorganic sulfate.

For the purpose of enzyme purification, the cells were grown with choline sulfate supplying all three elements and were harvested at late log or early stationary phase. The cell extract was fractionated with ammonium sulfate and the 0 to 50% saturation fraction collected and further fractionated on a DEAE-cellulose (DE-23) column. The overall purification achieved was about 12-fold with a recovery of 49%. This preparation (certainly still highly impure) was then used for a number of kinetic and other

Table 5
SOME PROPERTIES OF CHOLINE SULFATASES OF BACTERIAL AND FUNGAL ORIGIN

Enzyme source	Degree of purification (×)	Molecular weight	Optimum pH	K_m (mM)	Inhibition by				Type of enzyme
					SO_3^{2-}	Thiols	Thiol-group reactive reagents	Choline	
Bacteria									
Pseudomonas nitroreducens	4	—	8.3	40.0	Yes	—	—	Yes	Inducible
P. aeruginosa	12	175,000	7.5	1.4—2.5	—	Yes	?	Yes	Inducible
Pseudomonas 5A	0	—	—	—		—	—	—	Inducible
Klebsiella aerogenes	0	—	—	—	—	—	—	—	Constitutive
Fungi									
Penicillium chrysogenum	0	345,000	8.0	8.5	Yes	Yes	Yes	Yes	Constitutive
Aspergillus nidulans	60—100	430,000	8.2	12.5	Yes	Yes	Yes	Yes	Derepressible
Neurospora crassa	0	—	5.6, 7.8[a]	—	Yes	Yes[b]	Yes	Yes	Derepressible

[a] Two pH optima seen.
[b] But not by L-cysteine.

experiments. Some general properties are recorded in Table 5, but those of immediate interest include the molecular weight of 175,000 and the K_m varying between 1.4 and 2.5 m*M* (depending on the particular experiment) which was substantially lower than that found for the enzyme from *P. nitroreducens*.[76] The enzyme was subject to product inhibition by choline and was inactivated by 10^{-4} *M* dithiothreitol or 10^{-3} *M* mercaptoethanol. Some thiol group reactive reagents (iodoacetamide, iodoacetate) did not inhibit the enzyme, whereas others (*p*-chloromercuribenzoate, *N*-ethylmaleimide) did. The investigators proposed that the enzymic reaction proceeded by an ordered BiBi mechanism with water being the second substrate to bind to the enzyme and choline being the first product to leave. Base catalysis was envisaged, without the involvement of a covalently bound sulfate-enzyme intermediate. Further details of the proposed mechanism are provided in Volume II, Chapter 2.

3. The Choline Sulfatase of Aspergillus nidulans

It will be recalled from an earlier section that Spencer and his colleagues initially failed to detect choline sulfatase activity in *A. nidulans* (wild type and several mutants) because of the sulfur-sufficient state of the organism. In a subsequent study,[91] some properties of the enzyme in crude cell extracts were described. Again a high K_m was noted (35 m*M*) and the enzyme was particularly intolerant of temperatures in excess of 28° C. Complete and rapid loss in activity followed the addition of L-cysteine.

Further information about this enzyme does not appear to have been published, but some very interesting and detailed studies are reported in a Ph.D. thesis by Lee from the same laboratory.[92] In this work a mutant of *A. nidulans*, designated as S1 and lacking PAPS-reductase, was selected on the basis of preliminary growth and enzyme level studies. It was further shown that maximum enzyme production occurred when this mutant was grown for between 85 and 100 hr in a modified Czapek-Dox medium containing 0.05 m*M* sodium thiosulfate as sulfur source. Other preliminary work established that the enzyme was particularly labile at its optimum pH (8.2), but was relatively stable at pH 6.6.

Armed with this information, the investigator developed a purification procedure that involved absorption of the enzyme on to calcium phosphate gel at pH 6.0 and subsequent elution at pH 7.2, batchwise fractionation on DEAE-cellulose followed by fractionation on Sephadex® G-200. In different experiments a purification varying from 60- to 100-fold was achieved with recovery varying from about 20 to 50%. A coexisting arylsulfatase activity was eliminated at the final purification stage. Polyacrylamide gel electrophoresis of the best preparation revealed the presence of two contaminating proteins (one present in substantial quantities) in addition to the enzyme. A finding that will be relevant to any future investigator was that the purified enzyme preparation tended to lose activity by contact with glass. This could be prevented by precoating the glass with a film of albumin and the preparation could then be stored at −20° C (13% loss in activity in 5 weeks).

Some general properties of the enzyme are recorded in Table 5 and require particular mention. The K_m of the enzyme (12.5 m*M*) is still somewhat high although much less than that reported for the crude enzyme (35 m*M*).[91] The enzyme was confirmed as a simple hydrolase producing equimolar amounts of choline and sulfate. It did not hydrolyse 1-methylcholine sulfate, aminoethyl sulfate, dimethylaminoethyl sulfate, ethyl sulfate, D-glucose 6-sulfate, chondroitin sulfate, choline phosphate, or acetylcholine. The molecular weight of this fungal enzyme (430,000) was more than twice that of the enzyme from the bacterium *P. nitroreducens* (175,000),[76] but was not so markedly different from that reported (345,000) for a crude preparation of the choline sulfatase of another fungus, *Penicillium chrysogenum*.[90] A considerable number of potential inhibitors of the enzyme were examined. Thiol group reactive agents seemed to have little effect on the enzyme but, as was found for the choline sulfatase of *Pseudomonas nitroreducens*, sulfite was a potent inhibitor. In common with the *P. aeruginosa* enzyme there was product inhibition by choline and almost complete inhibition by compounds possessing thiol groups (L-cysteine, *N*-acetyl-L-cysteine, reduced glutathione, dithiothreitol, and mercaptoethanol). This last finding was perhaps the most significant of all and evidence was subsequently obtained indicating that the effect of a thiol compound such as L-cysteine or dithiothreitol was to dissociate much of the enzyme protein into eight identical but inactive subunits of 53,000 mol wt. Some inactive dimers also appear to be formed. Substantial reactivation of the enzyme could be achieved by treatment with L-dehydroascorbic acid. It was further concluded that the inhibition of the enzyme by sulfite was a reflection of inactivation caused by the rupture of disulfide bridges between subunits. Further comment on the effects of thiols and sulfite is made in a later section.

4. Other Choline Sulfatases

Although much work on the physiological and genetic control of their syntheses has been done, no enzymological studies of significance appear to have been reported for other known sources of choline sulfatase. Relatively little is therefore known about the bacterial enzymes from *K. aerogenes* or *Pseudomonas* V-A or the fungal enzymes of *N. crassa* and *Penicillium chrysogenum*. However, brief comments about the *P. chrysogenum* enzyme have been made,[85,90] indicating a K_m of 8.5 m*M*, inhibition by thiol group reactive reagents and by thiols, product inhibition by choline, and a molecular weight of about 345,000. Similarly, McGuire[93] has provided some brief details of the enzyme present in cell extracts of *N. crassa* that again pointed to the instability of the enzyme and to the pronounced inhibition by sulfite and high concentrations of mercaptoethanol or dithiothreitol (5 m*M* or greater). On the other hand, cysteine apparently activated this enzyme (60%) at a concentration (5 m*M*) that would have given considerable inactivation of the corresponding enzyme from *A. nidulans.*[92] *p*-Chloromercuribenzoate and cystine also inhibited, and in the latter case the inhibition could not be reversed by cysteine or dithiothreitol.

5. Significance of the Collective Results of Enzymological Studies

Table 5 summarizes some of the more important facts that have emerged from the very limited studies that have been made on microbial choline sulfatases. The bacterial enzymes studied so far have been either constitutive or inducible, while the fungal enzymes are either constitutive or derepressible. In no case can the affinity of enzyme for substrate be said to be great but, because K_m values were obtained with enzyme preparations that were generally highly impure, it would be unwise to place any undue physiological or environmental relevance on this. In all cases where a study has been made, the enzymes are product-inhibited by choline and this fact has possibly been ignored by some investigators who have chosen to measure the activity of the enzyme under arbitrary conditions.

Thiol group reactive reagents have inhibited enzyme activity in all the enzymes that have so far been examined and this may indicate the importance of one or more thiol groups to the catalytic activity of the enzymes. However, perhaps the most significant of all the findings concern the effects of sulfite and thiols on the enzymes. Although the information is far from complete and an anomaly exists in the shape of the activating effect of cysteine on the enzyme from *N. crassa*, it seems possible that further work on pure enzymes might indicate the sulfite and thiol effects to be a common feature of choline sulfatase enzymes. If so, then the studies on the *A. nidulans* enzyme[91,92] take on a particular significance and need to be extended to the enzymes from other sources. Thus, the work of Scott and Spencer[91] had shown that choline sulfatase could not be detected in mycelia grown on a sulfur-sufficient medium, but that subsequent sulfur deprivation led to the appearance of the enzyme. Moreover, L-cysteine inhibited the activity of the enzyme both in vivo and in vitro and it was suggested that the amino acid regulated the enzyme, first by repressing its synthesis, and second by inhibiting the activity of the enzyme. Sulfite was considered to share the last function. Subsequently, Lee[92] established that treatment with thiols or sulfite led to the dissociation (reversible) of the enzyme into inactive monomeric and dimeric subunits and postulated the involvement of the dissociation/reassociation phenomenon in the regulation of the enzyme. He proposed the following scheme to accommodate the various findings.

$$\text{Active choline sulfatase octamer} \underset{\substack{\text{Sulfur-deficiency} \\ \text{(cell in oxidized state)}}}{\overset{\substack{\text{Sulfur-sufficiency} \\ \text{(cell in reduced state)}}}{\rightleftarrows}} \text{Inactive monomers (and dimers)} \qquad (2)$$

It is first assumed that the widespread occurrence in quantity of choline sulfate in the higher fungi points to a role as a sulfur store built up in times of sulfur sufficiency and utilized in times of sulfur deficiency. The internal compensation of charge of the ester would allow it to act as an important sulfur store within the cell without problems of disturbed osmotic balance. At the molecular level under circumstances of sulfur sufficiency, L-cysteine and reduced glutathione (and possibly sulfite) accumulate in the cells of *A. nidulans* together with choline sulfate. The thiol compounds dissociate the active choline sulfatase enzyme molecules by rupturing disulfide bridges so that under these circumstances choline sulfate is stored. In times of sulfur insufficiency, accumulated L-cysteine and reduced glutathione are depleted as a result of protein synthesis, the inactive monomers and dimers of choline sulfatase recombine to form active octameric enzyme, and sulfate can then be released from the stored choline sulfate for assimilatory purposes. The precise mechanism for the reoxidation of the subunit is as yet unknown, although this is readily achieved in vitro with L-dehydroascorbate.

Of course these proposals have been based on a study of mycelial sources of the enzyme. However, Takebe[94] has found choline sulfate to be a major component of the conidiospores of *A. niger* and other *Aspergillus* species, while McGuire and Marzluf[95] have indicated that the metabolic regulation of choline sulfatase synthesis in conida of *N. crassa* differs from that operating in the mycelial stage of development. Moreover, although the bacterium *Lactobacillus plantarum* appears to contain choline sulfate as an intracellular component,[7] *Pseudomonas* species excrete the ester into the growth medium.[78] Choline sulfate does not, therefore, appear to be serving as a sulfur store for pseudomonads. Indeed, Gravel[96] has cast some doubts on the storage role of choline sulfate and has collated data that collectively argue against a primary role for the ester in sulfur storage. He has further suggested that choline sulfate may be involved in the regulation of the pool sizes of PAPS and its immediate precursor adenosine 5′-phosphosulfate (APS). It is envisaged that ability to synthesize choline sulfate would ensure that APS and PAPS did not accumulate to cellular levels high enough to interfere with the energy balance of the cell or to inhibit other important adenylate-dependent reactions. In this way choline sulfate production provides a route for the immobilization of the sulfate ion so that it does not become available for reactivation. In Gravel's view any mechanism for the hydrolysis of PAPS that regenerated inorganic sulfate would simply produce a never ending energy requiring cycle between sulfate and PAPS. He further suggested the possibility that excessive accumulation of APS and PAPS might be toxic to growth and points to the studies of Gillespie et al.[97] that indicated that the accumulation of these compounds in *Salmonella typhimurium* mutants, defective in ability to reduce PAPS, led to decreased cell viability.

Gravel's ideas are certainly interesting and would still hold good for bacteria of the *Pseudomonas* genus that are able to achieve the sulfation of choline by enzymes that appear to be cell-wall associated and are actually secreted into the culture fluid in considerable amounts. Again it could be argued that this mechanism enables surplus inorganic sulfate to be disposed of, hence reducing the possibility of high levels of APS and PAPS. Further comments on the involvement of APS and PAPS are made in Volume II, Chapter 4.

In any event, it now seems essential to extend the excellent studies initiated in Spencer's laboratory[91,92] in order to pursue further the various concepts and possibilities that seem to be emerging. In retrospect, it seems rather unfortunate that the results of the detailed work on *A. nidulans*[92] have not been made available to the general scientific community.

6. The Strange Case of Pseudomonas C12B

The subject of choline sulfatase cannot be left without some mention of the utilization of its substrate by *Pseudomonas* C12B. This bacterium was able to grow on choline sulfate as sole sulfur source and this suggested the involvement of a choline sulfatase.[98] However, cell extracts of the organism, while apparently able to liberate choline from ^{35}S-labeled choline sulfate, liberated only traces of $^{35}SO_4^{2-}$ ions. Similar findings were made under varying experimental conditions, and with extracts of cells harvested at different stages of the culture cycle during growth on 0.1 m*M* choline sulfate. Other experiments established that maximum production of the choline sulfate utilizing system occurred when citrate and L-cysteine were used as carbon/energy and sulfur sources, respectively. Under these circumstances most of the activity was released into the culture medium. Again choline was liberated from choline ^{35}S-sulfate by dialyzed culture medium, but no $^{35}SO_4^{2-}$ ions could be detected. Similar results were obtained when the $BaCl_2$ gelatin method was employed to estimate inorganic sulfate liberated from unlabeled choline sulfate. In other experiments[99] a sensitive microbiological assay procedure for detecting enzymically liberated sulfate also gave negative results. The investigators reluctantly concluded that a simple choline sulfatase was not involved in the utilization of choline sulfate by *Pseudomonas* C12B.

This mystery was compounded somewhat by the fact that, in a different study,[89] no difficulty had been experienced in detecting sulfate liberated from choline sulfate by a soil bacterium designated as *Pseudomonas* V-A. Moreover, it will be recalled that *Pseudomonas* C12B can produce five distinct simple alkylsulfatases, the activities of which can be followed with the $BaCl_2$ gelatin assay method for enzymically liberated sulfate. Of course, it is possible that a choline sulfatase is present in the organism but that the liberated sulfate is transferred directly to an acceptor other than water. However, the investigators[98] prefer to consider again the possibility raised earlier[80] that a choline sulfotransferase enzyme may be operating in reverse, transferring sulfate from choline sulfate to PAP. The thermodynamic problems associated with this suggestion still remain a formidable theoretical obstacle to this possibility.

For the moment, the mechanism associated with the utilization of choline sulfate by *Pseudomonas* C12B must remain a mystery that begs to be reinvestigated. Such an investigation should also take into consideration the fact that the bacterium also secretes into the growth medium the complete enzyme system for the biosynthesis of choline sulfate from ATP, choline, and inorganic sulfate. The relationship between the two systems is therefore of the greatest interest and may well provide some vital new information of relevance to bacterial sulfur metabolism.

IV. PHENOXYETHYL SULFATASE

The use of sodium 2-(2,4-dichlorophenoxy) ethyl sulfate (CRAG herbicide, Figure 7) as a herbicide was first described by King and Lambrech.[100] The compound can be used as a preemergence treatment for weed control and has the useful property that it is inherently inert and therefore does not injure any foliage receiving spray drifts. Because the compound is a sulfate ester (Figure 7) the action of the herbicide was assumed to be dependent on the ability of soil microorganisms to remove the ester sulfate group and convert the liberated 2-(2,4-dichlorophenoxy)ethanol to the active herbicide, 2,4-dichlorophenoxyacetic acid (2,4-D). Circumstantial evidence for this came from the finding the CRAG herbicide was not effective in sterilized soils[101] and from more direct evidence[102] that indicated the conversion of the herbicide to 2,4-D by unidentified soil microorganisms. Subsequently, in careful and detailed studies, Vlitos[103,104] showed that under soil pH conditions that precluded acid hydrolysis, the sulfate ester was converted to the corresponding free alcohol and 2,4-D. Moreover, the alcohol was active in its own right in regulating growth in cucumber roots and tomato plants. Various soil bacterial isolates were examined for their ability to "activate" the sulfated herbicide and a number of such "activators" were found. One particular organism, *Bacillus cereus* var. *mycoides* was selected for more detailed studies and was grown on nutrient broth or on a synthetic medium containing glucose as carbon and energy source and inorganic sulfate as sulfur source. After growth on either source it was possible to show that the cell-free culture media could "activate" the herbicide. Moreover, with the synthetic culture medium it was possible to measure the amount of 2-(2,4-dichlorophenoxy)ethanol liberated. No attempts were made to detect liberated inorganic sulfate and no arylsulfatase activity could be detected in either culture medium. This latter observation seemed to deter the investigator from a positive conclusion concerning the presence of a new sulfatase, although subsequent reviewers of the sulfatase field have tended to accept that some type of alkylsulfatase enzyme must be responsible. In this connection, it is noteworthy that 2-(2,4-dichlorophenoxy)ethyl sulfate can serve as an inducer of the P2 alkylsulfatase of *Pseudomonas* C12B.[47] However, attempts to detect sulfatase activity toward the ester by that enzyme or by crude cell extracts were unsuccessful and the bacterium failed to grow when the ester was used as sulfur source (unpublished results from the Cardiff laboratories).

The unsatisfactory state of affairs in relation to the "activation" of CRAG herbicide has led to a reinvestigation of the problem in Cardiff. At the present time nothing has been published, but some comments on progress might be of interest to the reader.

Enrichment culture with the herbicide serving as a sulfur source has led to the isolation of five different bacteria that can liberate 2-(2,4-dichlorophenoxy)ethanol from the corresponding sulfate. One of these organisms (still not fully identified, but designated as *Pseudomonas* FL) has been selected for further studies. It appears that the enzyme may be constitutive as it is produced by broth-grown and pyruvate-grown cells in the absence of the sulfate ester; moreover, there appears to be no requirement for the ester when maintaining the organism on agar slants. Some progress has been made at the enzymological level. The activity of the enzyme toward the substrate is assayed by measuring the liberated parent alcohol, while the position of the enzyme on gel electrophoretograms is determined by incubating gels with substrate in the presence of barium acetate. A white band of precipitated $BaSO_4$ appears at the position taken up by the enzyme. The optimum pH of the enzyme is in the region of 6.0 and product inhibition of enzyme activity means that some care has to be taken when measuring activity for kinetic purposes. Maximum sulfatase activity in cells is obtained in the late stationary phase of growth, just before appreciable cell lysis can be observed.

This work is continuing but, meanwhile, it is pertinent to note that the organism appears also to have the capacity to produce three simple secondary alkylsulfatases and three simple primary alkylsulfatases when grown on nutrient broth or on a basal salts medium containing pyruvate. Preliminary gel electrophoresis experiments indicate that one of the secondary enzymes is active toward D-decan-2-yl sulfate, but not to the L-form of that ester but, at the time of writing, it is not clear (because of their close proximity to each other on gels) whether either of the other two secondary alkylsulfatases is stereospecific for the L-form. In any event, these preliminary findings indicate the *Pseudomonas* FL to produce more alkylsulfatases than the much studied *Pseudomonas* C12B, and this without the need to include specific inducers in the growth medium. Of particular interest is that one of the three primary alkylsulfatase bands overlaps completely with the enzyme band that hydrolyses 2-(2,4-dichlorophenoxy)ethyl sulfate, so that it is possible that a single enzyme is responsible for both activities.

Finally, some detergent formulations contain long-chain alkylethoxy sulfate esters. Studies currently in progress in Cardiff have shown that the biodegradation of one such ester, dodecyltriethoxy sulfate, when use to supply carbon and energy for the growth of a pseudomonad designated as DES-1, is initiated principally by cleavage of an ether bond(s) in the triethoxy moiety rather than by the action of a sulfatase. Cell extracts of the organism grown in this way appear to be inactive in hydrolysing CRAG herbicide (which also contains an ethoxy sulfate group). It will now be interesting to see whether dodecyltriethoxy sulfate can serve as a substrate for the phenoxyethyl sulfatase of *Pseudomonas* FL. Whether or not there are naturally occurring substrates for that enzyme will have to await separation of the enzyme from the other coexisting sulfatases, followed by detailed specificity studies. Meanwhile, it is interesting to speculate as to whether the chlorinated alkyl diol disulfate of *Ochromonas danica* or the 2,3-dibromo-5-hydroxybenzyl-1′,4-disulfate of the red alga, *Rhodomela larix* (see Section II.) might eventually emerge as natural substrates for the enzyme.

V. STEROL SULFATASE — LITHOCHOLATE SULFATASE

The existence of enzymes that could hydrolyse sterol sulfate esters was first indicated by Henri et al.,[105] who noted that the intestinal juice of the terrestial mollusk, *Helix pomatia*, was able to hydrolyse dehydroepiandrosterone sulfate (Figure 8). A similar enzyme was later found in other mollusks and in mammalian tissues.[106] This type of sulfatase appears to be relatively highly specific in hydrolysing only the 3β-sulfates of the Δ^5 series of sterols. It was therefore also active toward such compounds as cholesterol sulfate (Figure 8) and epiandrosterone sulfate. Considerable interest is currently being shown in the mammalian enzyme and in its various substrates; first, because it is now clear that both have important physiological roles to play with respect to the transformation of steroids[107] and, second, because of two clinical conditions (multiple sulfatase deficiency and placental sulfatase deficiency) in which the enzyme is missing.[108]

Whether other types of mammalian sterol sulfatases occur that act on the α-sulfate group or on ester sulfate groups attached directly to other ring carbons of the sterol nucleus is still not absolutely clear, but it seems probable from the work of Iwamori et al.,[109] that there is a distinct enzyme that is able to hydrolyse testosterone sulfate (C-17 sulfate group, Figure 8).

In relation to the possibility of similar sulfatase enzymes being present in microorganisms, it is interesting to recall that the search for such enzymes in the early 1950s was principally stimulated by the need to find appropriately mild methods of hydrolysing steroid conjugates in order that the chemical structure of the liberated sterol could be preserved intact for subsequent assay. At the time there was no indication that the

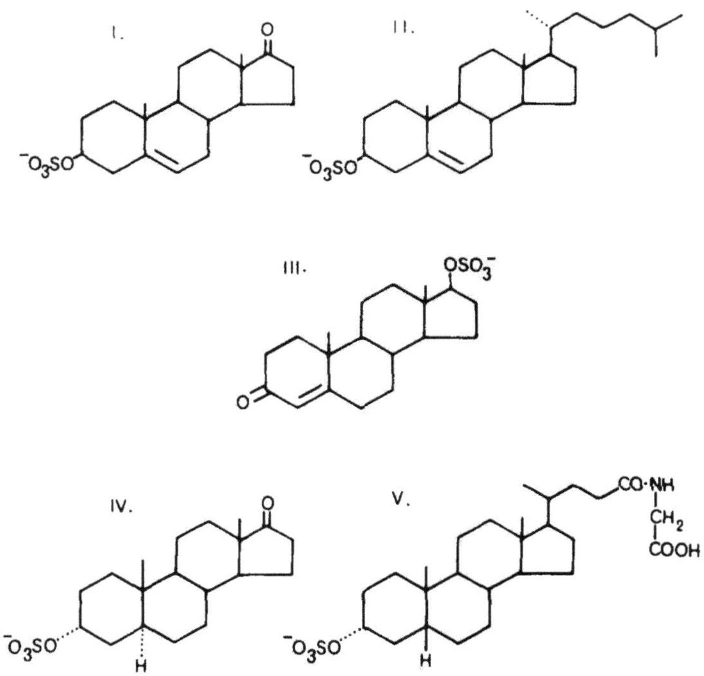

FIGURE 8. Examples of naturally occurring potential substrates for microbial sterol sulfatases. I, dehydroepiandrosterone 3β-sulfate; II, cholesterol 3β-sulfate; III, testosterone 17-sulfate; IV, androsterone 3α-sulfate; V, glycolithocholate 3α-sulfate.

substrates were not necessarily metabolic end-products or that the enzymes might have any physiological role to play. It was thus particularly discouraging when a search for a microbial source of the enzyme proved fruitless.[110] Of bacteria, 23 types were examined as possible sources of an "alcoholic" sulfatase, but sulfatase activity of any type was found in only two of them *(Proteus vulgaris* and *Pseudomonas non liquifaciens).* Growing cultures and the culture fluids of these two organisms could liberate sulfate from chondroitin sulfate, but did not hydrolyse the sulfate esters of cholic acid (3α-sulfate), androsterone (3α-sulfate), or dehydroepiandrosterone (3β-sulfate). Moreover, the crude commercial arylsulfatase-containing fungal preparations, Taka Diastase® and Mylase® P, were without activity toward the last two substrates. Almost 20 years were to elapse before any further serious attempts were made to resume the search for bacterial sterol sulfatases.

A. Potential Substrates

Potential substrates for bacterial sterol sulfatases are numerous and varied and no attempt is made here to provide an exhaustive list. However, the following remarks will perhaps illustrate the variety of compounds that might eventually reach the domain of microorganisms in soils and waters and in the digestive systems of animal organisms.

An indication of the physiological importance of mammalian 3β-sterol sulfatase and its substrates has already been given and, elaborating on this point, it is now clear that certain of the substrates can take part in metabolic pathways that operate in parallel to those followed by the parent steroids. For example, cholesterol sulfate can follow, without loss of the sulfate group, the metabolic pathway to dehydroepiandrosterone sulfate and then eventually to estriol.[107,111] Moreover, the relationship between this pathway in the fetoplacental unit and placental 3β-sterol sulfatase is underlined by the fact that absence of the enzyme in placental sulfatase deficiency leads to a failure of the increase in estrogen production observed in normal pregnancy and consequently to pregnancy complications.[108] Cholesterol sulfate is also a stabilizing component of the mammalian erythrocyte membrane,[112] while another 3β-sulfate, desmosterol sulfate, apparently plays a similar stabilizing role with respect to the spermatozoan membrane. It is suggested that 3β-sterol sulfatase, which is present throughout the female reproductive tract and which exhibits cyclical activity, may release the sulfate moiety and hence be responsible for the destabilization of the spermatozoan membrane during capacitation.[113] These various mammalian 3β-sulfate esters, together with several others not described, can therefore be presumed to be potential substrates for microbial enzymes. Many lower animals also produce sterol sulfates, for example cholesterol sulfate has been found in echinoderms and starfish,[114] while the 3β-sulfate esters of campesterol and β-sitosterol, as well as cholesterol, are excretory products of the tobacco hookworm, *Manduca sexta*.[115] It is highly probable that searches would reveal the ubiquitous distribution of this type of compound in the animal world.

Other sterol sulfates in which the sulfate group is esterified to a 3α-hydroxyl group are found in excreta and tissue fluids of mammals and are generally regarded as metabolic end-products.[116] Examples include androsterone sulfate (Figure 8) and etiocholanolone sulfate. Other steroids possessing sulfate groups attached directly to the D-ring at position 17 also occur in tissue fluids or urine (e.g., testosterone sulfate, Figure 8). A new twist to the story of mammalian sterol sulfates came with the discovery[117,118] that the 3α-hydroxyl group of each of the bile acids, glycolithocholate (Figure 8), and taurolithocholate, was readily sulfated by mammalian organisms. Between 40 and 80% of the lithocholate content of human bile was present in the form of these 3α-sulfate esters. Other forms of sulfated bile salts also occur in mammalian bile and excreta; for example, chenodeoxycholate 7-sulfate and cholate 7-sulfate have been detected in the intestines and feces of both convential and germ-free mice.[119] It will be seen in the next section that the discovery of sulfated bile salts was an important factor in the renewal of interest in microbial sterol sulfatases.

Further potential naturally occurring substrates for the enzymes include the myxinol disulfate (one 3β-sulfate group) of hagfish bile (Figure 3) and the complex sulfated sterol saponins that serve as protection agents in members of the starfish family.[120-123] One of these, thornasteroside A, a major component of the toxic sterol saponin complement of the starfish *Acanthaster planci*, has been fully characterized (Figure 9) and contains a 3β-sulfate ester group, while the sterol moiety of a minor component (thornasteroside B), not yet fully characterized, also possesses a 3β-sulfate group.[122] There is also a possibility that corynoline 11-sulfate (Figure 9), the first sulfated alkaloid to be found[124] in a higher plant *(Corydalis incisa)*, might also be a potential substrate for a sterol sulfatase enzyme.

With such a variety of naturally occurring sterol sulfates, most of which might be supposed to return to soils and waters, it would indeed be surprising if sterol sulfatase enzymes did not occur in microorganisms of one sort or another.

B. The Discovery of Microbial Sterol Sulfatases

The lack of interest that followed the initial search for these enzymes[110] was dispelled in part by fundamental studies on bacterial metabolism in the gastrointestinal tracts of germ-free rats, but particularly by the finding[125] that bile acid metabolism in humans could involve the formation of sulfate esters. In the latter context, space does not permit a detailed account of subsequent developments but, briefly, the work from Palmer's laboratory[117,118,125] that pinpointed the occurrence of sulfated glyco- and taurolithocholates in human bile led to an immediate interest on the part of clinical inves-

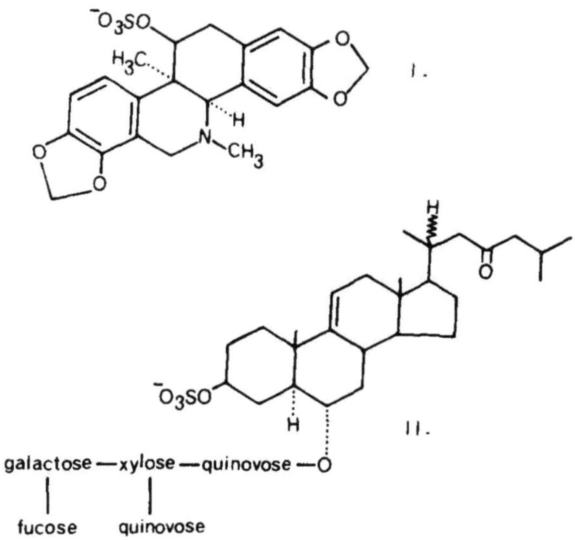

FIGURE 9. Examples of naturally occurring potential substrates for microbial sterol sulfatases. I, corynoline 11-sulfate; II, thornasteroside A.

tigators. The reason for this stemmed from the knowledge that lithocholate and its amino acid conjugates can cause pronounced liver damage.[126] In normal subjects the levels of sulfated and nonsulfated lithocholate conjugates are relatively low, but increase markedly in various disease conditions including obstruction of the bile duct, hepatitis, and cirrhosis (see Reference 12 for brief review). Furthermore, there now exists a growing use of orally administered chenodeoxycholate for the dissolution of gallstones and, because this bile acid is readily converted to lithocholate by intestinal microorganisms, the long-term use of such treatment is open to question. Man appears to deal with the problem of lithocholate toxicity by sulfating the 3α-hydroxyl group of either the free acid or the corresponding amino acid conjugates. This not only leads to decreased hepatoxicity,[12] but also allows a reduction in the amount of the circulating lithocholate moiety by enabling urine to become a major route of its excretion.[127] Clearly, if intestinal microorganisms are able to remove the ester sulfate group from the lithocholate moiety, then the potential danger to the patients remains, particularly in cases where relatively high concentrations of the ester are being presented to the gut flora over a long period of time, hence increasing the possibility of sulfatase "induction". The findings (see Reference 128 for literature citations) that lithocholate can enhance the mutagenicity of certain known carcinogens gives further emphasis to the danger.

In fact the general interest in the ability of intestinal microorganisms to modify bile acids and other sterols had been the subject of many investigations before the discovery of sulfated bile acids,[129,130] and the first excursions into the sulfatase field really arose because of the growing interest in the use of germ-free rats for experimental purposes. Thus Eriksson and Gustafsson[62] showed that the proportion of free steroids in the feces of germ-free rats was much smaller than that present in feces from conventional rats. These investigators demonstrated that ester sulfate groups present at the 3α-, 3β-,17β-, and 21-hydroxyl groups of sterol moieties could all be hydrolysed by rat caecal bacteria. The substrates used in these experiments included the 3β-sulfate esters of dehydroepiandrosterone and cholesterol, the 3α-sulfate ester of 3α-hydroxy-5α-pregnan-20-one, and the 17β-sulfate ester of testosterone. Substrates were incubated with caecal contents of conventional and germ-free rats and desulfation of all the substrates was observed in the former animals, but not the latter. No attempts were made to isolate any microorganisms that might have been responsible for the observed desulfation; nevertheless the work provided firm indications of sterol sulfatase activity of microbial origin.

Subsequent work has been concerned with bile salt sulfate esters and in spite of the encouraging results of the experiments with germ-free and conventional rats, the first investigations by Bremmelgaard and Bremmelgaard[131] were singularly disappointing. More than 100 strains, representing 26 species of bacteria generally found in the human intestinal tract, were tested for their ability to metabolize the 3α-sulfate ester of taurolithocholate. The organisms were grown on a complex serum-containing medium and although taurolithocholate sulfate was converted to lithocholate sulfate by strains of *P. aeuruginosa, Streptococcus faecalis, Bacteroides* species and *Clostridium welchii,* in no case was there any indication of sulfatase activity.

Fortunately, this setback did not deter further studies of the problem and Imperato and his colleagues[132] were soon able to provide the stimulus that will now hopefully lead to further studies. These investigators isolated a strain of *P. aeruginosa* from human feces and cultured the organism on a minimal support medium containing ^{35}S-labeled lithocholate sulfate as sulfur source. No hydrolysis of the ester occurred over the first 6 hr, but by the 12th hr 50% of the substrate had been hydrolysed, and after 24 hr all the substrate had been hydrolysed. It was concluded that the optimal time for enzyme induction was about 13 hr. Significantly, cell growth on occasions was not accompanied by hydrolysis of the sulfate ester and this led the investigators to examine the possibility that inorganic sulfate was serving to repress enzyme production. This indeed proved to be the case and very little enzyme activity could be detected in cells that had been grown in the usual medium but with the addition of 50 m*M* magnesium sulfate. No similar sterol sulfatase activity could be detected in *Escherichia coli, B. fragilis, B. cadaveris, S. faecalis,* and bifidobacteria. The investigators indicated that isolation and characterization of the *P. aeruginosa* enzyme was under way but, as yet, no further report seems to have appeared.

Meanwhile, Kelsey et al.[128] have examined the ability of human mixed fecal microorganisms (grown anaerobically) to metabolize lithocholate sulfate. Significant conversion of the ester to sulfated cholenates (Δ^2 and Δ^3) was observed, as well as a relatively low rate of hydrolysis of the parent ester to the free bile acid. Only weak sulfatase activity was therefore being exhibited.

This study and that of Bremmelgaard and Bremmelgaard,[131] when considered in the light of the inhibition by sulfate of sterol sulfatase production,[132] perhaps suggests that failure to detect the enzyme (or the detection of low levels of activity) may reflect the growth conditions employed. This is clearly a subject that cries out for further study. At the same time, however, it must be appreciated that the growth conditions employed in any in vitro experiments may be far removed from those existing in vivo, and this pinpoints the difficulty of applying to the human situation the important metabolic techniques that have been applied to conventional and germ-free experimental animals.

Meanwhile, only the surface of the interesting field of microbial sterol sulfatases has been scratched and much remains to invite attention. At the worst, sulfatase enzymes may be isolated that could play a useful role in the analysis of steroids, while at the best, information may emerge that could be of considerable relevance to clinical problems associated with the gut and the enterohepatic circulation.

REFERENCES

1. **Harada, T.**, The formation of sulphatases by *Pseudomonas aeruginosa*, *Biochim. Biophys. Acta*, 81, 193, 1964.
2. **Tudball, N. and O'Neill, J. G.**, The utilization of L-serine O-sulphate by *Pseudomonas* FR, *Microbios*, 13, 217, 1975.
3. **Tudball, N. and O'Neill, J. G.**, Substrate specificity of the L-serine-O-sulphate degrading activities of *Pseudomonas* FR, *Biochim. Biophys. Acta*, 429, 616, 1976.
4. **Dodgson, K. S. and Tudball, N.**, Enzymic desulphation of L-serine O-sulphate: the intracellular localization of the enzyme, *Biochem. J.*, 81, 68, 1961.
5. **Tudball, N., Thomas, J. H., and Fowler, J. A.**, A rat liver system that catalyses a pyridoxal phosphate-independent $\alpha\beta$-elimination, *Biochem. J.*, 114, 299, 1969.
6. **Haines, T. H.**, Halogen- and sulfur-containing lipids of *Ochromonas*, *Ann. Rev. Microbiol.*, 27, 403, 1973.
7. **Mercer, E. J. and Davies, C. L.**, Distribution of chlorosulpholipids in algae, *Phytochemistry*, 18, 457, 1979.
8. **Payne, W. J. and Painter, B. G.**, Resolution by acrylamide gel electrophoresis of alkyl sulphatases and alcohol dehydrogenase, *Microbios*, 3, 199, 1971.
9. **Liem, P. Q. and Laur, M-H.**, Les alcools aliphatiques sulfatés: nouveaux lipides polaires isolés de diverse fucacées, *Biochimie*, 58, 1381, 1976.
10. **Weinstein, B., Rold, T. L., Harrell, C. E., Burns, M. W., and Waaland, J. R.**, Re-examination of the bromophenols of the red alga, *Rhodomela larix*, *Phytochemistry*, 14, 2667, 1975.
11. **Hubel, W. and Hahrstedt, A.**, Cardiospermine sulfate — a sulphur-containing cyanogenic glucoside from *Cardiospermum grandiflorum*, *Tetrahedron Letts.*, No. 45, 4395, 1979.
12. **Haslewood, G. A. D.**, *The Biological Importance of Bile Salts*, North-Holland, Amsterdam, 1978, chap. 5.
13. **Yagi, T.**, Sulfate esters in developing eggs, *J. Biochem. Tokyo*, 59, 495, 1966.
14. **Lee, M. L., Later, D. W., Rollins, D. K., Eatough, D. J., and Hansen, L. D.**, Dimethyl and monomethyl sulfate: presence in coal fly ash and airborne particulate matter, *Science*, 207, 186, 1980.
15. **Solomon, S. and Bhavnani, B. R.**, The isolation and metabolism of conjugates of neutral steroids from natural sources, in *Chemical and Biological Aspects of Steroid Conjugation*, Bernstein, S. and Solomon, S., Eds., Springer-Verlag, Berlin, 1970, 321.
16. **Pasqualini, J. R.**, Metabolic conjugation and hydrolysis of steroid hormones in the feto-placental unit, in *Metabolic Conjugation and Metabolic Hydrolysis*, Vol. 2, Fishman, W. H., Ed., Academic Press, New York, 1970, 153.
17. **Williams, O. B. and Rees, H. B.**, Bacterial utilization of anionic surface-active agents, *J. Bacteriol.*, 58, 823, 1949.
18. **Payne, W. J. and Feisal, V. E.**, Bacterial utilization of dodecyl sulfate and dodecyl benzene sulfonate, *Appl. Microbiol.*, 11, 339, 1963.
19. **Hsu, Y-C.**, Detergent (sodium lauryl sulphate)-splitting enzyme from bacteria, *Nature (London)*, 200, 1091, 1963.
20. **Payne, W. J.**, Pure culture studies of the degradation of detergent compounds, *Biotechnol. Bioeng.*, 5, 355, 1963.
21. **Williams, J. and Payne, W. J.**, Enzymes induced in a bacterium by growth on sodium dodecyl sulfate, *Appl. Microbiol.*, 12, 360 1964.
22. **Hsu, Y-C.**, Detergent-splitting enzyme from *Pseudomonas*, *Nature (London)*, 207, 385, 1965.
23. **Payne, W. J., Williams, J. P., and Mayberry, W. R.**, Primary alcohol sulfatase in a *Pseudomonas* species, *Appl. Microbiol.*, 13, 698, 1965.
24. **Williams, J. P., Mayberry, W. R., and Payne, W. J.**, Metabolism of linear alcohols with various chain lengths by a *Pseudomonas* species, *Appl. Microbiol.*, 14, 156, 1966.
25. **Payne, W. J. and Painter, B. G.**, Resolution by acrylamide gel electrophoresis of alkyl sulphatases and alcohol dehydrogenase, *Microbios*, 3, 199, 1971.
26. **Payne, W. J., Williams, J. P., and Mayberry, W. R.**, Hydrolysis of secondary alcohol sulphate by a bacterial enzyme, *Nature (London)*, 214, 623, 1967.
27. **Payne, W. J., Fitzgerald, J. W., and Dodgson, K. S.**, Methods for visualization of enzymes in polyacrylamide gels, *Appl. Microbiol.*, 27, 154, 1974.
28. **Fitzgerald, J. W. and Payne, W. J.**, Induction in a *Pseudomonas* species of sulphatases active on short chain alkylsulphates, *Microbios*, 5, 87, 1972.
29. **Dodgson, K. S., Fitzgerald, J. W., and Payne, W. J.**, Chemically defined inducers of alkylsulphatases present in *Pseudomonas* $C_{12}B$, *Biochem. J.*, 138, 53, 1974.
30. **Fitzgerald, J. W.**, Hydrolysis of sodium hexan-1-yl sulphate by the P2 electrophoretic form of primary alkylsulphatase, *Microbios*, 11, 153, 1974.

31. **Fitzgerald, J. W., Dodgson, K. S., and Matcham, G. W. J.**, Secondary alkylsulphatases in a strain of *Comamonas terrigena, Biochem. J.*, 149, 477, 1975.
32. **Matcham, G. W. J., Dodgson, K. S., and Fitzgerald, J. W.**, Purification properties and cellular localization of the stereospecific CS2 secondary alkylsulphohydrolase of *Comamonas terrigena, Biochem. J.*, 167, 723, 1977.
33. **Matcham, G. W. J. and Dodgson, K. S.**, Preparation and characterization of substrates suitable for the study of stereospecific secondary alkylsulphohydrolases of detergent-degrading micro-organisms, *Biochem. J.*, 167, 717, 1977.
34. **Bartholomew, B., Dodgson, K. S., Matcham, G. W. J., Shaw, D. J., and White, G. F.**, A novel mechanism of enzymic ester hydrolysis: inversion of configuration and carbon–oxygen bond cleavage by secondary alkylsulphohydrolases from detergent-degrading micro-organisms, *Biochem. J.*, 165, 575, 1977.
35. **Barrett, C. H., Dodgson, K. S., and White, G. F.**, Further studies on substrate and inhibitor specificity of the stereospecific CS2 secondary alkylsulphohydrolase of the detergent-degrading microorganism *Comamonas terrigena, Biochem J.*, 191, 467, 1980.
36. **Spencer, B.**, Enzymic cleavage of aryl hydrogen sulphates in presence of $H_2{}^{18}O$, *Biochem. J.*, 69, 155, 1958.
37. **Spencer, B.**, Sulphatases XXV. Determination of $BaS^{16}O_3{}^{18}O$ by infrared spectroscopy, *Biochem. J.*, 73, 442, 1959.
38. **Matcham, G. W. J.**, Studies on Detergent-Degrading Enzymes from *Comamonas terrigena*, Ph.D. thesis, University of Wales, Cardiff, Wales, 1976.
39. **Bartholomew, B., Dodgson, K. S., and Gorham, S. D.**, Purification and properties of the S1 secondary alkylsulphohydrolase of the detergent-degrading micro-organism, *Pseudomonas* $C_{12}B$, *Biochem. J.*, 169, 717, 1978.
40. **Matcham, G. W. J., Bartholomew, B., Dodgson, K. S., Fitzgerald, J. W., and Payne, W. J.**, Stereospecificity and complexity of microbial sulphohydrolases involved in the biodegradation of secondary alkylsulphate detergents, *FEMS Microbiol. Letts.*, 1, 197, 1977.
41. **Fitzgerald, J. W. and Laslie, W.W.**, Loss of primary alkylsulfatase and secondary alkylsulfatases (S1 and S2) from *Pseudomonas* $C_{12}B$: effect of culture conditions, cell-washing procedures, and osmotic shock, *Can. J. Microbiol.*, 21, 59, 1975.
42. **Mandrand-Berthelot, M-A., Novel, G., and Novel, M.**, L'induction gratuite de la β-glucuronidase d' *Escherichia coli* K12 et son double mécanisme de répression, *Biochimie*, 59, 163, 1977.
43. **Blackwell, C. M. and Turner, J. M.**, Microbial metabolism of amino alcohols: formation of coenzyme B_{12}-dependent ethanolamine ammonia-lyase and its concerted induction in *Escherichia coli, Biochem. J.*, 176, 751, 1978.
44. **Shaw, D. J., Dodgson, K. S., and White, G. F.**, Substrate specificity and other properties of the inducible S3 secondary alkylsulfohydrolase purified from the detergent-degrading bacterium *Pseudomonas* $C_{12}B$, *Biochem. J.*, 187, 181, 1980.
45. **Dodgson, K. S., Fitzgerald, J. W., and Payne, W. J.**, Localization of the S3 secondary alkylsulfohydrolase in *Pseudomonas* $C_{12}B$, *FEMS Microbiol. Letts.*, 3, 115 1978.
46. **Fitzgerald, J. W., Dodgson, K. S., and Payne, W. J.**, Induction of primary alkylsulphatases and metabolism of sodium hexan 1-yl sulphate by *Pseudomonas* $C_{12}B$, *Biochem. J.*, 138, 63, 1974.
47. **Cloves, J. M., Dodgson, K. S., White, G. F., and Fitzgerald, J. W.**, Specificity of P2 primary alkylsulphohydrolase induction in the detergent-degrading bacterium *Pseudomonas* $C_{12}B$, *Biochem. J.*, 185, 13, 1980.
48. **Cloves, J. M., Dodgson, K. S., White, G. F., and Fitzgerald, J. W.**, Purification and properties of the P2 primary alkylsulphohydrolase of the detergent-degrading bacterium *Pseudomonas* $C_{12}B$, *Biochem. J.*, 185, 23, 1980.
49. **Zwann, J.**, Estimation of molecular weights of proteins by polyacrylamide gel electrophoresis, *Anal. Biochem.*, 21, 155, 1967.
50. **Cloves, J. M., Dodgson, K. S., Games, D. E., Shaw, D. J., and White, G. F.**, The mechanism of action of primary alkylsulfohydrolase from a detergent-degrading micro-organism, *Biochem. J.*, 167, 843, 1977.
51. **Swisher, R. D.**, *Surfactant Biodegradation*, Marcel Dekker, New York, 1970.
52. **Flawn, P. C., Malcolm, S. A., and Woodroffe, R. C. S.**, Assessment of the preservative capacity of shampoos, *J. Soc. Cosmet. Chem.*, 24, 229, 1973.
53. **Rotmistrov, M. N., Stavskaya, S. S., Krivetz, I. A., and Samoylenko, L. S.**, Microorganisms decomposing alkyl sulfates, *Microbiology*, 47, 338, 1978.
54. **Kiprianova, E. A., Stavskaya, S. S., Boiko, O. I., and Krivetz, I. A.**, Assimilation of sodium dodecyl sulfate by bacteria of the *Pseudomonas* genus, *Microbiol. Zh. (Kiev)*, 40, 503, 1978.
55. **Fitzgerald, J. W. and Kight, L. C.**, Physiological control of alkylsulfatase synthesis in *Pseudomonas aeruginosa:* effect of glucose, glucose analogs and sulfur, *Can. J. Microbiol.*, 23, 1456, 1977.

56. **Kight-Oliff, L. C. and Fitzgerald, J. W.**, Inhibition of enzyme induction in *Pseudomonas aeruginosa* by exogenous nucleotides, *Can. J. Microbiol.*, 24, 811, 1978.
57. **Fitzgerald, J. W., Kight-Oliff, L. C., Stewart, G. J., and Beauchamp, N. F.**, Reversal of succinate-mediated catabolite repression of alkylsulfatase in *Pseudomonas aeruginosa* by 2,4-dinitrophenol and by sodium malonate, *Can. J. Microbiol.*, 24, 1567, 1978.
58. **Stavskaya, S. S., Udilova, O. F., and Rotmistrov, M. N.**, Study of bacterial alkylsulfatase localization, *Dopov. Acad. Nauk, Ukr. RSR. Ser. B.*, 12, 1118, 1978.
59. **Sukhumavasi, J., Murooka, Y., and Harada, T.**, Formation of sulfatases in *Klebsiella aerogenes, J. Ferment. Technol.*, 53, 62, 1975.
60. **Lijmbach, G. W. M. and Brinkhuis, E.**, Microbial degradation of secondary *n*-alkyl sulfates and secondary alkanols, *Antonie van Leeuwenhoek, J. Microbiol. Serol.*, 39, 415, 1973.
61. **Vaicum, L. and Eminovici, A.**, Contributions to the knowledge of the hydrolytic biodegradation mechanism of alkyl sulfate detergents, *Rev. Roum. Biochim.*, 2, 149, 1976.
62. **Eriksson, H. and Gustafsson, J.-A.**, Steroid in germfree and conventional rats: sulpho- and glucuronohydrolase activities of caecal contents from conventional rats, *Eur. J. Biochem.*, 13, 198, 1970.
63. **Barrett, C. H., Dodgson, K. S., White, G. F., and Payne, W. J.**, Preliminary studies on dehydrogenases in *Comamonas terrigena* that exhibit stereospecificity towards secondary alcohols, *Biochem. J.*, 187, 703, 1980.
64. **Schmidt, E. and Wagner, W.**, Ueber Cholin, Neurin und Verwandte verbindungen, *Justus Liebigs Ann. Chem.*, 337, 51, 1904.
65. **Woolley, D. W. and Peterson, W. H.**, The chemistry of mold tissue. XIV. Isolation of cyclic choline sulfate from *Aspergillus sydowi, J. Biol. Chem.*, 122, 213, 1937.
66. **Raistrick, H. and Vincent, J. M.**, A survey of fungal metabolism of inorganic sulphates, *Biochem. J.*, 43, 90, 1948.
67. **Hockenhull, D. J. D.**, Studies in penicillin production by *Penicillium notatum* in surface culture. II. Further studies in the metabolism of sulfur, *Biochem. J.*, 43, 498, 1948.
68. **De Flines, J.**, The occurrence of a sulfuric acid ester of choline in the mycelium of a strain of *Penicillium chrysogenum, J. Am. Chem. Soc.*, 77, 1676, 1955.
69. **Stevens, C. M. and Vohra, P.**, Occurrence of choline sulfate in *Penicillium chrysogenum, J. Am. Chem. Soc.*, 77, 4935, 1955.
70. **Lindberg, B.**, Studies on the chemistry of lichens. VIII. Investigation of a *Dermatocarpon* and some *Rocella* species, *Acta Chem. Scand.*, 9, 917, 1955.
71. **Lindberg, B.**, Methylated taurines and choline sulphate in red algae, *Acta Chem. Scand.*, 9, 1323, 1955.
72. **Nissen, P. and Benson, A. A.**, Choline sulfate in higher plants, *Science*, 134, 1759, 1961.
73. **Egami, F. and Itashi, M.**, Biochemical studies on choline sulfate ester. I. *Aspergillus oryzae* grows in a medium with choline sulfate ester as sulfur source, *Igaku Seibutsugaku*, 19, 292, 1951.
74. **Harada, T. and Spencer, B.**, Choline sulphate in fungi, *J. Gen. Microbiol.*, 22, 520, 1960.
75. **Itashi, M.**, Comparative biochemistry of choline sulfate metabolism, *J. Biochem. Tokyo*, 50, 52, 1961.
76. **Takebe, I.**, Isolation and characterization of a new enzyme choline sulfatase, *J. Biochem. Tokyo*, 50, 245, 1961.
77. **Ikawa, M., Chakravarti, A., and Taylor, R. F.**, Occurrence of choline in *Lactobacillus plantarum, Can. J. Microbiol.*, 18, 1241, 1972.
78. **Fitzgerald, J. W.**, Formation of choline *O*-sulphate by *Pseudomonas* C12B and other *Pseudomonas* species, *Biochem. J.*, 136, 361, 1973.
79. **Kaji, A. and Gregory, J. D.**, Mechanism of sulfurylation of choline, *J. Biol. Chem.*, 234, 3007, 1959.
80. **Spencer, B. and Harada, T.**, The role of choline sulfate in the sulfur metabolism of fungi, *Biochem. J.*, 77, 305, 1960.
81. **Orsi, B. A. and Spencer, B.**, Choline sulphokinase (sulphotransferase), *J. Biochem. Tokyo*, 56, 81, 1964.
82. **Renosto, F. and Segel, J. H.**, Choline sulfokinase of *Penicillium chrysogenum:* partial purification and kinetic mechanism, *Arch. Biochem. Biophys.*, 180, 416, 1977.
83. **Fitzgerald, J. W. and Luschinski, P. C.**, Further studies on the formation of choline sulfate by bacteria, *Can. J. Microbiol.*, 23, 483, 1977.
84. **Gregory, J. D. and Lipmann, F.**, The transfer of sulfate among phenolic compounds with 3′, 5′-diphosphoadenosine as coenzyme, *J. Biol. Chem.*, 229, 1081, 1957.
85. **Segel, I. H. and Johnson, M. J.**, Hydrolysis of choline-*O*-sulfate by cell-free extracts from *Penicillium, Biochim, Biophys. Acta*, 69, 433, 1963.
86. **Harada, T.**, The formation of sulphatases by *Pseudomonas aeruginosa, Biochim. Biophys. Acta*, 81, 193, 1964.
87. **Spencer, B., Hussey, E. C., Orsi, B. A., and Scott, J. M.**, Mechanism of choline-*O*-sulphate utilization in fungi, *Biochem. J.*, 106, 461, 1968.

88. **Sukhumavasi, J., Murooka, Y., and Harada, T.**, Formation of sulfatases in *Klebsiella aerogenes, J. Ferment. Technol.*, 53, 62, 1975.
89. **Fitzgerald, J. W. and Scott, C. L.**, Utilization of choline O-sulphate as a sulfur source for growth by a *Pseudomonas* isolate, *Microbios*, 10, 121, 1974.
90. **Lucas, J. J., Burchiel, S. W. and Segel, I. H.**, Choline sulfatase of *Pseudomonas aeruginosa, Arch. Biochem. Biophys.*, 153, 664, 1972.
91. **Scott, J. M. and Spencer, B.**, Regulation of choline sulfatase synthesis and activity in *Aspergillus nidulans, Biochem. J.*, 106, 471, 1968.
92. **Lee, Y. S.**, Choline Sulphatase, Ph.D. thesis, Trinity College, University of Dublin, Ireland, 1977.
93. **McGuire, W. G.**, Sulfur Storage in *Neurospora crassa:* Regulation of Choline O-Sulfate Metabolism in Several Developmental Stages, Ph.D. thesis, Ohio State University, Columbus, Ohio, 1973.
94. **Takebe, I.**, Choline sulfate as a major soluble sulfur component of conidiospores of *Aspergillus niger, J. Gen. Appl. Microbiol.*, 6, 83, 1960.
95. **McGuire, W. G. and Marzluf, G. A.**, Developmental regulation of choline sulfatase and aryl sulfatase in *Neurospora crassa, Arch. Biochem. Biophys.*, 161, 360, 1974.
96. **Gravel, R. A.**, Choline O-sulfate utilization in *Aspergillus nidulans, Genet. Res. Camb.*, 28, 261, 1976.
97. **Gillespie, D., Demerec, M., and Itakawa, H.**, Appearance of double mutants in aged cultures of *Salmonella typhimurium* cysteine-requiring strains, *Genetics*, 59, 433, 1968.
98. **Fitzgerald, J. W., Luschinski, P. C., and Scott, C. L.**, Utilization of choline sulphate for growth by a choline sulphohydrolase-deficient *Pseudomonas* species, *Int. J. Biochem.*, 9, 343, 1978.
99. **Fitzgerald, J. W., Cline, M. E., and Rose, F. A.**, Sulfate esters as derepressors of arylsulphatase formation and the problem of sulphate ester utilization by *Pseudomonas* C12B, *FEMS Microbiol. Lett.*, 2, 217, 1977.
100. **King, L. J. and Lambrech, J. A.**, A new chemical for use in pre-emergence weeding, Proc. Northeast Weed Control Conf., New York City, N.Y., January 1949, p. 34.
101. **King, L. J., Lambrech, J. A., and Finn, T. P.**, Herbicidal properties of sodium 2,4-dichlorophenoxyethyl sulfate, *Contrib. Boyce Thompson Inst.*, 16, 191, 1950.
102. **Audus, J. L.**, Fate of sodium 2,4-dichlorophenoxyethyl-sulphate in the soil, *Nature (London)*, 170, 886, 1952.
103. **Vlitos, A. J.**, Biological activation of sodium 2-(2,4-dichlorophenoxy) ethyl sulfate by *Bacillus cereus* var. *mycoides, Contrib. Boyce Thompson Inst.*, 16, 435, 1952.
104. **Vlitos, A. J.**, Biological activation of sodium 2-(2,4-dichlorophenoxy) ethyl sulfate, *Contrib. Boyce Thompson Inst.*, 17, 127, 1953.
105. **Henri, R., Thevenet, M., and Jarrige, P.**, The hydrolysing action of glucuronidase and phenolsulfatase of the digestive juice of *Helix pomatia, Bull. Soc. Chim. Biol.*, 34, 897, 1952.
106. **Dodgson, K. S. and Rose, F. A.**, Sulfohydrolases, in *Metabolic Pathways*, Vol. 7, *Metabolism of Sulfur Compounds*, Greenberg, D. M., Ed., Academic Press, New York, 1975, chap. 9.
107. **Gower, D. B. and Fotherby, K.**, Biosynthesis of the androgens and oestrogens, in *Biochemistry of Steroid Hormones*, Makin, H. L. J., Ed., Blackwell Scientific, Oxford, 1975, chap. 4.
108. **Farooqui, A. A.**, Sulfatases, sulfate esters and their metabolic disorders, *Clin. Chim. Acta*, 100, 285, 1980.
109. **Iwamori, M., Moses, H. W., and Kishimoto, Y.**, Steroid sulfatase in brain: comparison of sulfohydrolase activities for various steroid sulfates in normal and pathological brains, including the various forms of metachromatic leukodystrophy, *J. Neurochem.*, 27, 1389, 1976.
110. **Buehler, H. J., Katzmann, P. A., and Doisy, E. A.**, Hydrolysis of conjugates of neutral ketosteroids, *Proc. Soc. Exp. Biol. Med.*, 78, 3, 1951.
111. **Roberts, K. D. and Lieberman, S.**, The biochemistry of the 3β-hydroxy-Δ^5-steroid sulfates, in *Chemical and Biological Aspects of Sulfate Conjugation*, Bernstein, S. and Solomon, S., Eds., Springer-Verlag, Berlin, 1970, chap. 5.
112. **Bleau, G., Lauiere, G., Chapdelaine, A., and Roberts, K. D.**, Red cell surface structure, stabilization by cholesterol sulfate as evidenced by scanning electron microscopy, *Biochim. Biophys. Acta*, 375, 220, 1975.
113. **Legault, Y., Bouthillier, M., Bleau, G., Chapdelaine, A., and Roberts, K. D.**, The sterol and sterol sulfate content of the male hamster reproductive tract, *Biol. Reprod.*, 20, 1213, 1979.
114. **Goad, L. J.**, The steroids of marine algae and invertebrate animals, in *Biochemical and Biophysical Perspectives in Marine Biology*, Mahns, D. C. and Sargent, J. R., Eds., Vol. 3, Academic Press, New York, 1976, 213.
115. **Hutchins, R. F. N. and Kaplains, J. N.**, Sterol sulfates in an insect, *Steroids*, 13, 605, 1969.
116. **Hadd, H. E. and Blickenstaff, R. T.**, *Conjugates of Steroid Hormones*, Academic Press, New York, 1969, chaps. 2 and 6.
117. **Palmer, R. H. and Bolt, M. G.**, Bile acid sulfates. I. Synthesis of lithocholic acid sulfates and their identification in human bile, *J. Lipid Res.*, 12, 680, 1971.

118. **Palmer, R. H.,** Bile acid sulfates. II. Formation, metabolism and excretion of lithocholic acid sulfates in the rat, *J. Lipid Res.*, 12, 680, 1971.
119. **Eyssen, H. J., Parmentier, G. G. and Mertens, J. A.,** Sulfated bile acids in germ-free and conventional mice, *Eur. J. Biochem.*, 66, 507, 1976.
120. **Mackie, A. M. and Turner, A. B.,** Partial characterization of a biologically active steroid glycoside isolated from the starfish, *Marthasterids glacialis*, *Biochem. J.*, 117, 543, 1970.
121. **Kitagawa, I., Kabayashi, M., and Sugawara, T.,** Steroidal saponins from the starfish, *Acanthaster planci* L (Crown of Thorns). I. Structures of two genuine sapogenols, thornasterol A and thornasterol B and their sulfates, *Chem. Pharm. Bull.*, 26, 1852, 1978.
122. **Kitagawa, I. and Kobayashi, M.,** Steroidal saponins from the starfish *Acanthaster planci* L (Crown of Thorns). II. Structure of the major saponin thornasteroside A, *Chem. Pharm. Bull.*, 26, 1864, 1978.
123. **Schmitz, F. J.,** Uncommon marine steroids, in *Marine Natural Products Chemical and Biological Perspectives*, Vol. 1, Scheuer, P. J., Ed., Academic Press, New York, 1978, chap. 5.
124. **Isawa, K., Takas, N., Nonaka, G., and Nishioka, I.,** (+)-Corynoline 11-*O*-sulfate from *Corydalis incisa*, *Phytochemistry*, 18, 1725, 1979.
125. **Palmer, R. H.,** The formation of bile acid sulfates: a new pathway of bile acid metabolism in humans, *Proc. Nat. Acad. Sci. U.S.A.*, 58, 1047, 1967.
126. **Palmer, R. H.,** Bile acids, liver injury and liver disease, *Arch. Intern. Med.*, 130, 606, 1972.
127. **Van Berge Henegouwen, G. P., Brandt, K-H., Eyssen, H., and Parmentier, G.,** Sulfated and unsulfated bile acids in serum, bile, and urine of patients with cholestasis, *Gut.*, 17, 861, 1976.
128. **Kelsey, M. I., Muschick, G. M., and Sexton, S. A.,** The metabolism of lithocholic acid- 3α-sulfate by human intestinal microflora, *Lipids*, 13, 152, 1978.
129. **Hayakawa, S.,** Microbiological transformations of bile acids, *Adv. Lipid Res.*, 11, 143, 1973.
130. **Shimada, K., Bricknell, K. S., and Finegold, S. M.,** Deconjugation of bile acids by intestinal bacteria: review of literature and additional studies, *J. Infect. Dis.*, 119, 273, 1969.
131. **Bremmelgaard, A. and Bremmelgaard, A.,** Bacterial metabolization of taurolithocholic acid 3-α-sulfate, *Acta Path. Microbiol. Scand., Sect. B.*, 82, 537, 1974.
132. **Imperato, T. J., Wong, C. G., Chen, L. J., and Bolt, R. J.,** Hydrolysis of lithocholate sulfate by *Pseudomonas aeruginosa*, *J. Bacteriol.*, 130, 545, 1977.

Chapter 3

CARBOHYDRATE SULFATASES

I. INTRODUCTION

It is virtually certain that several sulfatase enzymes active toward sulfated carbohydrate residues are still awaiting discovery, and this reflects the relative lack of interest in what is perhaps the most challenging of all the research avenues presented by the microbial sulfatase enzymes. A bewildering conglomeration of carbohydrate sulfate esters exist in nature and probably constitute by far the largest proportion of the sulfur that returns to the environment in the form of sulfate esters. Some indication of the chemical variety of the naturally occurring sulfated carbohydrates will shortly be given and this variety needs to be viewed in the light of the mere handful of microbial carbohydrate sulfatase enzymes that have so far been studied. It is difficult to understand why this should have been such a neglected field, but one is tempted to identify two possible reasons. First, many of the potential substrates are difficult to isolate in pure form from natural sources; moreover many of them are polydisperse and have chemical structures that are still not precisely defined. Second, many of them are heteropolymers, the biodegradation of which requires the participation of several enzymes including sulfatases. Indeed, in a few cases it has been established that the action of a sulfatase is dependent on the prior activity of other enzymes that must first play their roles in depolymerizing the substrate. It shall be seen that similar considerations might eventually hold good for some nonpolymeric naturally occurring carbohydrate sulfates.

In discussing carbohydrate sulfatases it will be convenient to regard glycerol as a carbohydrate and the reasons for this will emerge subsequently. Meanwhile, so little is known about the precise specificity of those enzymes listed by the Enzyme Nomenclature Committee and others not so listed, that the existence as independent entities of some of the enzymes to be described is not yet clearly established. Eventually, it will be necessary to test each carbohydrate sulfatase for ability to act on a whole spectrum of carbohydrate sulfate esters before the position clarifies. The following list of microbial carbohydrate sulfatases is therefore to be regarded as tentative and serving as a guide to the further studies that now need to be made.

1. Glycosulfatase (sugar sulfate sulfohydrolase, EC 3.1.6.3), acting on sulfate esters of D-glucose and D-galactose and possibly other sulfated mono- and disaccharides.
2. Gluconate 6-sulfatase (not listed by the Enzyme Nomenclature Committee), acting on D-gluconate 6-sulfate.
3. Glycerate-sulfatase (not listed by the Enzyme Nomenclature Committee), acting on D-glycerate 3-sulfate.
4. κ-Carrageenan 4-sulfatase (not listed by the Enzyme Nomenclature Committee), acting on the sulfated neocarrabiose and neocarratetraose moieties resulting from the degradation of κ-carrageenan by κ-carrageenase.
5. Chondro-4-sulfatase (Δ,4,5-β-D-glucuronosyl-(1,4)-2-acetamido-2-deoxy-D-galactose-4-sulfate 4-sulfohydrolase, EC 3.1.6.9), acting on the ester sulfate group at C-4 of the *N*-acetyl-D-galactosamine moieties of saturated and unsaturated disaccharides resulting from the depolymerization of chondroitin 4-sulfate or dermatan 4-sulfate.
6. Chondro-6-sulfatase (Δ,4,5-β-D-glucuronosyl-(1,4)-2-acetamido-2-deoxy-D-galactose-6-sulfate 6-sulfohydrolase, EC 3.1.6.10), acting on the ester sulfate group

at C-6 of the *N*-acetyl-D-galactosamine moieties of saturated and unsaturated disaccharides resulting from the depolymerization of chondroitin 6-sulfate. Also acts on D-galactosamine 4,5-disulfate.

7. Keratan sulfatases (not listed by the Enzyme Nomenclature Committee), acting on keratan sulfate (containing *N*-acetyl-D-glucosamine 6-sulfate and D-galactose 6-sulfate residues). At least two sulfatases are involved, one acting on the sulfated hexosamine moieties and the other on the sulfated hexose residues.
8. Δ4,5,-Uronate sulfatase (not listed by the Enzyme Nomenclature Committee), acting on the C-2-sulfate groups of the uronic acid moiety of a trisulfated Δ4,5,-unsaturated disaccharide resulting from the depolymerization of heparan (and heparan sulfate).
9. Disulfoglucosamine 6-sulfatase (2-sulfamido-2-deoxy-6-*O*-sulfo-D-glucose 6-sulfohydrolase, EC 3.1.6.11), acting on D-glucosamine 2,6-disulfate and on D-glucosamine 6-sulfate, both products of the depolymerization of heparin (and heparan sulfate).
10. Sulfoglucosamine sulfamidase (2-sulfamido-2-deoxy-D-glucose sulfamidase, EC 3.10.1.1) acting on D-glucosamine 2,6-disulfate and on D-glucosamine *N*-sulfate, both products of the depolymerization of heparin (and heparan sulfate).

At first sight this would seem to be a formidable array of enzymes, but in fact all the information on them is contained in a handful of papers and in many cases the work has involved the use of crude enzyme preparations.

A. Potential Substrates

The potential substrates for carbohydrate sulfatases could themselves readily form the basis of a detailed monograph and, indeed, the compounds have already been the subject of an American Chemical Society Symposium.[1] Little more can therefore be done here than to indicate the great variety of potential substrates and the reader will be referred to reviews that will provide further information. Among these potential substrates are compounds in which ester sulfate groups are present on D-glucose, D- or L-galactose, 3,6-anhydro-D-galactose, L-rhamnose, L-fucose, D-mannose, D-xylose, D-arabinose, L-ascorbic acid, D-glucosamine, D-galactosamine, L-iduronic acid, and glycerol. The position of sulfate groups on an individual monosaccharide can also be variable. For example, ester sulfate groups have been noted on the 2-, 3-, 4-, and 6-positions of galactose in the sulfated polysaccharides of green seaweeds, while more than one sulfate is sometimes present on a single monosaccharide unit. In addition to sulfated substituted monosaccharides, sulfated substituted disaccharides, trisaccharides, tetrasaccharides, and polysaccharides also occur naturally, while in a number of cases the sulfated carbohydrate moieties are covalently associated with protein.

1. Potential Substrates Containing a Sulfated Monosaccharide Residue

Those of animal origin include ascorbate 2-sulfate found in the tissues and tissue fluids of mammals and other animal organisms;[2] UDP-*N*-acetyl-D-galactosamine 6-sulfate and the corresponding 4,6-disulfate, both isolated from avian oviducts;[3] cerebroside 3-sulfate (Figure 1, I) present in brain and kidney[4] and seminolipid 3-sulfate (Figure 1, II) present in mammalian testes and spermatozoa.[5] In the plant world are found the 6′-sulfate esters (position of sulfate presumed) of *p*-coumarylglucose (Figure 2, I) and caffeylglucose present in various species of fern,[6] anthroquinone glycoside sulfates such as emodin 1- (or 8-)glucose sulfate (Figure 2, II) found in the leaves of *Rumex pulcher*;[7] betanin 6′-sulfate (Figure 2, III) from the beetroot *Beta vulgaris*;[8] betanin 3-sulfate from the fruits of *Rivinia humalis*[9] and three of the four sulfolipids isolated by Liem and Laur[10] from brown seaweeds. These are, respectively, a 1,2-dia-

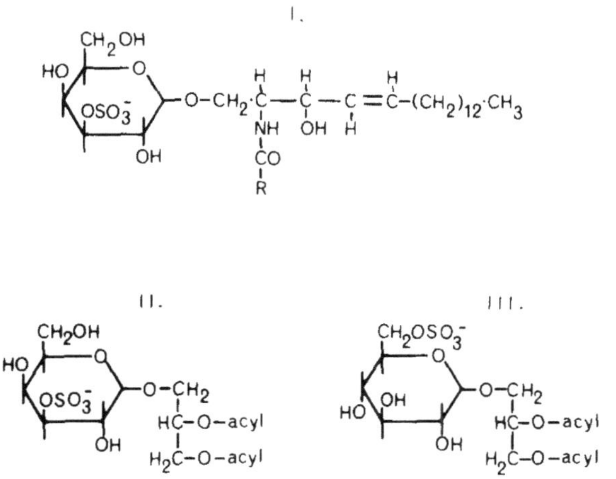

FIGURE 1. Examples of naturally occurring potential substrates for microbial carbohydrate sulfatases. Compounds containing a sulfated monosaccharide moiety. I, cerebroside 3′-sulfate; II, seminolipid 3′-sulfate; III, sulfolipid from brown seaweeds. See text for further details and origins.

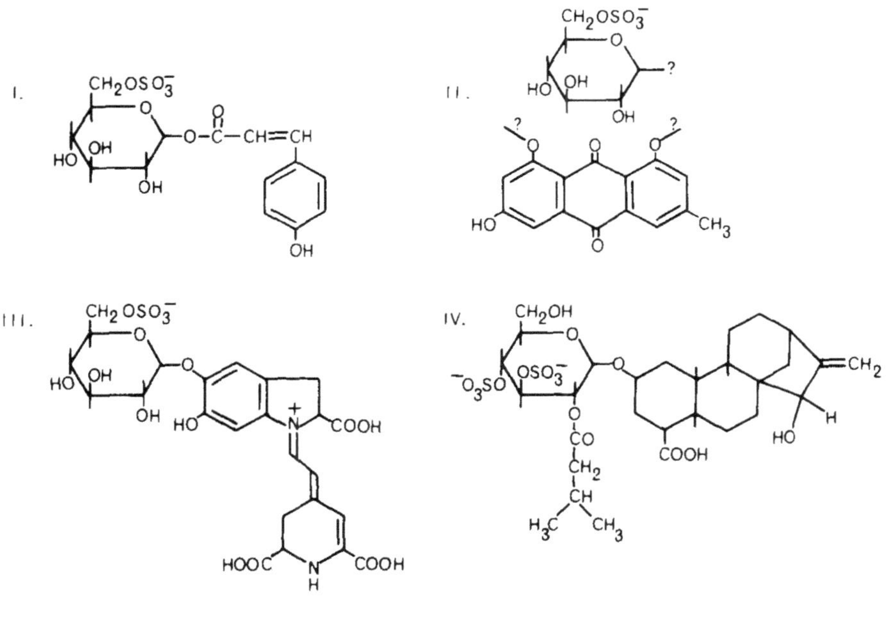

FIGURE 2. Examples of naturally occurring potential substrates for microbial carbohydrate sulfatases. Compounds containing a sulfated monosaccharide moiety. I, *p*-coumarylglucose 6′-sulfate; II, emodin 1- (or 8-)glucose 6′-sulfate; III, betanin 6′-sulfate; IV, atractyloside. See text for further details and origins.

cyl-3-(6′-sulfatoglucosyl) glycerol (Figure 1, III) 1,2-diacyl-3-(3′-sulfatoglucosyl) glycerol and 1,2-diacyl-3-(6′-sulfatogalactosyl) glycerol. An interesting toxic sulfated glycoside, atractyloside, which inhibits oxidative phosphorylation at very low doses, has been isolated from *Atractylis gummifera* and carries ester sulfate groups at C-3 and C-6 of its glucose moiety (Figure 2, IV).[11] Finally, and accepting glycerol as a mono-

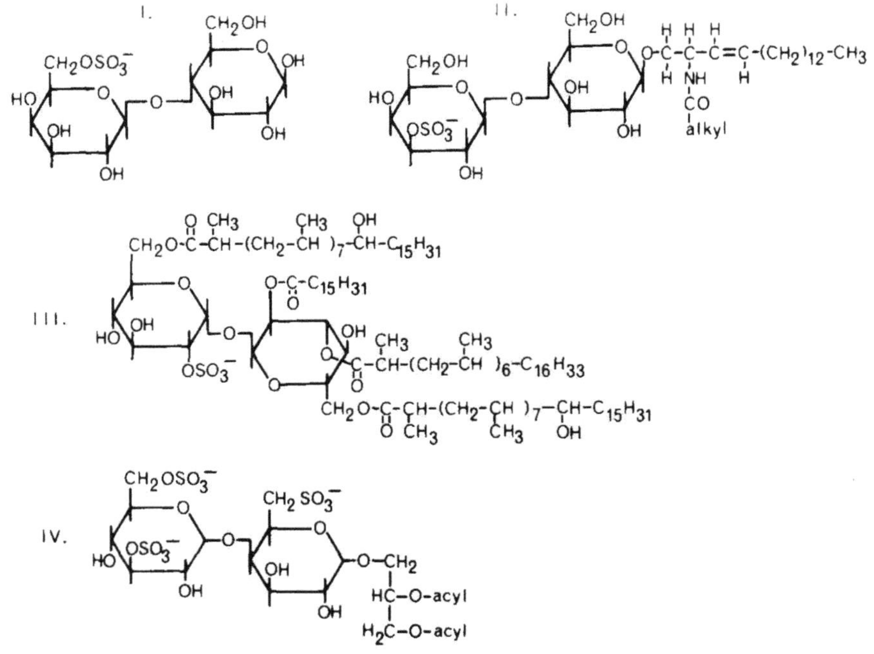

FIGURE 3. Examples of naturally occurring potential substrates for microbial carbohydrate sulfatases. Compounds containing a sulfated disaccharide moiety. I, lactose 6′-sulfate; II, lactosyl ceramide 3′-sulfate; III, sulfolipid from *Mycobacterium tuberculosis;* IV, disulfo-sulfonolipid from brown seaweeds. See text for further details and origins.

saccharide for the purpose of this discourse, an unusual phosphosulfolipid, 2,3-di*O*-(3′R,7′R,11′R,15-tetramethylhexadecyl)-*sn*-glycero-1-phosphoryl-3″-*sn*-glycerol-1″-sulfate, containing a sulfated glycerol residue, has been isolated from the extreme halophilic bacterium, *Halobacterium cutirubrum.*[12]

2. Potential Substrates Containing a Sulfated Disaccharide Moiety

At least three compounds possessing sulfated lactose residues are known, to be present in mammalian tissues. Lactose 6-sulfate[13] (Figure 3, I) and neuramin lactose 6-sulfate[14] have been isolated from mammary glands and lactosyl ceramide 3-sulfate (Figure 3, II) from human kidney and testis.[15,16] In the microbial world a family of sulfolipids containing trehalose 2′-sulfate moieties is present in virulent strains of *Mycobacterium tuberculosis* and appear to play an important role in the intracellular survival of this bacterium by antagonizing phagosomal-lysosomal fusion in phagocytic cells.[17] The most important of these sulfolipids is a 2,3,6,6′-tetraacyl-α,α'-D-trehalose 2′-sulfate (Figure 3, III) Finally, the brown seaweeds contain a sulfolipid in which a 1→4-linked, Glc-Gal disaccharide carries ester sulfate groups at C-3 and C-6 of the former sugar and a sulfonate group at C-6 of the latter (Figure 3, IV).[10]

3. Potential Substrates Containing a Sulfated Trisaccharide Moiety

Three such compounds have so far been found; the triglycosyl ceramide 3-sulfate[18] from mammalian gastric mucosa and containing a Gal-3-sulfate (1→4)- Gal-(1→4)-Glc trisaccharide component (Figure 4, I), the alkyl/acyl glycerol triglucosyl sulfate[19] from mammalian gastric secretion, with the terminal glucose being sulfated at C-6 (Figure 4, II), and the sulfolipid from *H. cutirubrum* containing a Gal-3-sulfate (1→6)-Man-(1→2)-Glc trisaccharide component (Figure 4, III).[20]

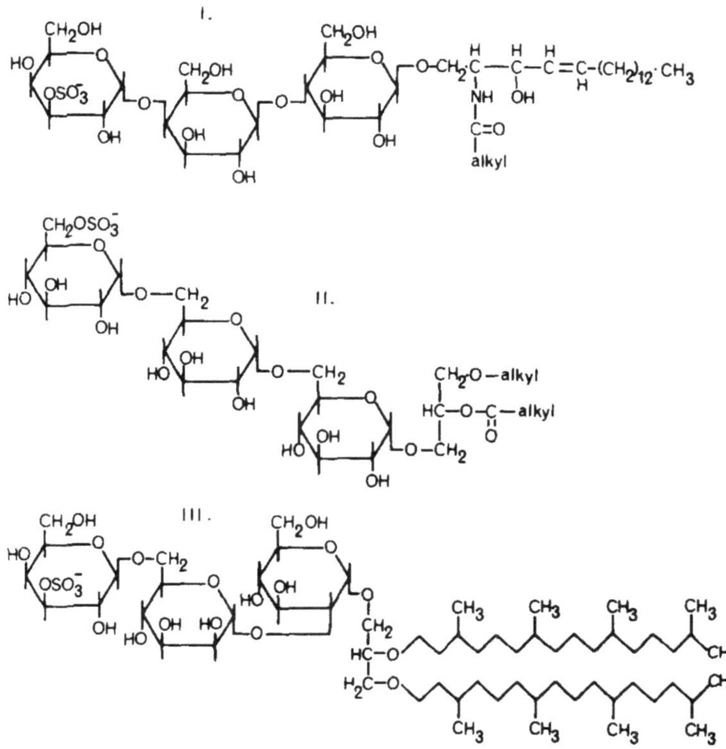

FIGURE 4. Examples of naturally occurring potential substrates for microbial carbohydrate sulfatases. Compounds containing a sulfated trisaccharide moiety. I, sulfolipid from gastric mucosa; II, sulfolipid from gastric secretions; III, sulfolipid from *Halobacterium cutirubrum*. See text for further details.

4. Potential Substrates Containing a Sulfated Tetrasaccharide Moiety

A major acidic glycolipid component of the surfactant of the rabbit lung has been identified[21] as an alkyl/acyl glycerol derivative possessing a sulfated tetrasaccharide component, Glc-6-sulfate (1→6)-Glc-(1→6)-Glc-(1→6)-Glc (Figure 5, I). Two interesting toxic chemicals (holothurins A and B) produced by the Holothurides (sea cucumbers) as a defense against predators, contain sulfated carbohydrate residues attached to steroid moieties.[22,23] In the case of holothurin A current information indicates a tetrasaccharide component of Xyl-sulfate-Glc-3-*O*-MethylGlc-Quin (Figure 5, II).[24]

Many of the compounds detailed in Sections A. 1. to 4 are sulfolipids and some general information relating to them has been collated in Table 1.

5. Potential Polysaccharide Substrates

Sulfated homo- and heteropolysaccharides probably account for the greatest proportion of the carbohydrate sulfates that eventually make their way back into the environment. With the exception of the marine algae, such sulfated polymers are particularly concentrated in the animal world, although examples do occur in microorganisms and in higher plants. Space permits no more than a brief and sometimes oversimplified account of these compounds and the reader is referred to detailed reviews for further information. In providing this account it will be convenient to subdivide these sulfated polymers into four different groups, namely, the polysaccharide sulfates of marine algae, polysaccharide sulfates of invertebrate animals, the sulfated glycosaminoglycans of animal connective tissues, and a miscellaneous group of polysaccharide sulfates from other sources.

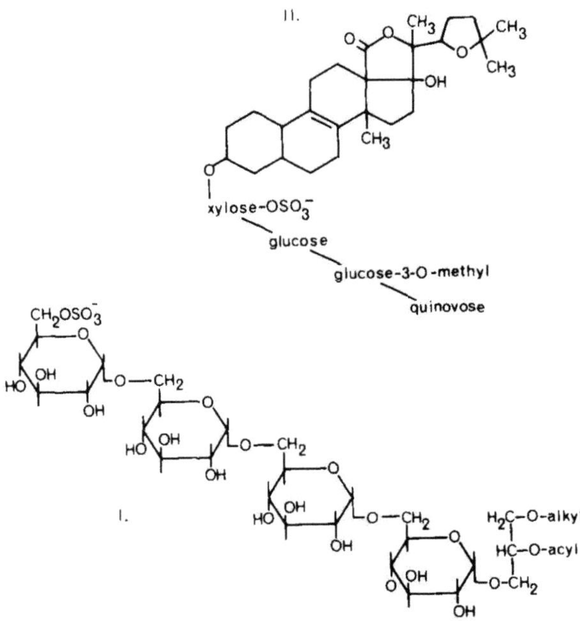

FIGURE 5. Examples of naturally occurring potential substrates for microbial carbohydrate sulfatases. Compounds containing a sulfated tetrasaccharide moiety. I, sulfolipid from rabbit lung surfactant; II, holothurin A. See text for further details.

a. Polysaccharide Sulfates of Marine Algae

Polysaccharides containing sulfated monomer units appear to have been found in most of the macroscopic marine algae that have been studied and there is growing evidence that similar compounds are an important feature of some unicellular algae that thrive in an aquatic environment. Frequently the concentration of ester sulfate groups is very high and the compounds constitute an extremely complex group of highly anionic polymers that can bind (and possibly exchange) cations, retain large amounts of water, and adopt conformations that may be presumed to be of value to the maintenance of algal structure, all properties that are relevant to survival in a harsh marine environment.

Members of the Phaeophyceae (brown seaweeds) possess fucose-containing sulfated polysaccharides that appear to serve as matrix polymers for the fibrillar polymers of the cell wall. Although generally referred to as fucans, they are really a family of polydisperse heteropolysaccharides containing varying amounts of galactose, mannose, xylose, and glucuronic acid in addition to the major component, fucose. Fractionation of extracts of different genera of brown seaweeds have yielded polymers containing, at one end of the spectrum, as much as 20% uronic acid and proportionally much less fucose and sulfate, while at the other end, 2 or 3% uronic acid and much higher quantities of fucose and sulfate. All are complex and highly branched and there is evidence for 1→2, 1→3, and 1→4 linked sulfated fucose units. In most cases the fucose units are sulfated at C-4, but in other polymers the sulfate group is on C-3. An excellent account of these compounds has been provided by Medcalf.[25]

The major sulfated polysaccharides of the Rhodophyceae (red seaweeds) are galactans composed of sulfated galactose or modified galactose units. Three types of polymer are found among the various members of the genera. First, a low sulfate, agar-

Table 1
COMPOSITION OF SOME NATURALLY OCCURRING SULFOLIPIDS

Trivial name	Source	Carbohydrate moiety	Aglycone	Ref.
Cerebroside sulfate	Brain	Gal 3-sulfate	Ceramide (acyl-sphingosine)	4
Seminolipid sulfate	Testes	Gal 3-sulfate	Diacylglycerol	5
Fucus sulfolipids[a]				10
1.	Brown seaweeds	Glc 6-sulfate	Diacylglycerol	
2.		Glc 3-sulfate	Diacylglycerol	
3.		Gal 6-sulfate	Diacylglycerol	
4.		Glc 3,6-disulfate, Gal 6-sulfonate	Diacylglycerol	
Lactosylceramide sulfate	Kidney	Gal 3-sulfate, Glc	Ceramide	16
Phytanylsulfolipid	*Halobacterium cutirubrum*	Gal 3-sulfate, man, Glc	Diphytanylglycerol	20
Mycobacterial sulfolipid	*Mycobacterium tuberculosis*	Trehalose 2′-sulfate	2,3,6,6′-tetracyl	17
Triglycosylceramide sulfate	Hog gastric mucosa	Gal 3-sulfate, Gal, Glc	Ceramide	18
Glyceroglucolipid sulfate	Human gastric secretion	Glc 6-sulfate, Glc, Glc	Alkyl, acylglycerol	19
	Rabbit lung	Glc 6-sulfate, Glc, Glc, Glc	Alkyl, acylglycerol	21

[a] Structures not completely defined.

type containing D- and L-galactose units as well as 3,6-anhydro-D-galactose. Secondly, a high sulfate carrageenan-type consisting of alternating units which can be designated A and B. In some carrageenans the A unit is D-galactose 4-sulfate and this may alternate with B units of D-galactose 6-sulfate, D-galactose 2, 6-disulfate, 3,6-anhydro-D-galactose 2-sulfate or 3,6-anhydro-D-galactose. Third, a mixed type of galactan exhibiting features of both the agarose and carrageenan-type of polymer. These various sulfated polysaccharides exist in vivo as mucilages and gels, the different physical properties of which reflect the differences in chemical structure and degree of sulfation. A useful review of these compounds has recently appeared.[26]

Before leaving the red seaweeds, a brief mention should be made of the production of extracellular polysaccharide sulfates by microscopic red algae of marine or fresh water origins such as *Rhodella maculata, Porphyridium cruentum* and *P. aerugineum.* The polymers provide a water-retaining gelatinous matrix in which the embedded cells can survive. In the case of the *Porphyridium* species the polymers appear to be covalently linked to protein.[27,28] The polysaccharide moieties are extremely complex, for example, the mucilage from the freshwater alga *P. aerugineum* contains D-xylose, D-glucose, D- and L-galactose, 3-*O*-methylxylose, 3- and 4-*O*-methylgalactose, D-glucuronic acid and 2,4-di-*O*-methylgalactose in the proportions 1.7:1.0:1.1:0.3:0.6:0.5:0.2. Ester sulfate groups account for 10% of the composition of the compound and are probably located on D-glucose and/or D-galactose residues.

Turning now to the Chlorophyceae (green seaweeds) we again encounter a variety of complex, polydisperse, sulfated heteropolysaccharides. All are highly branched but, unlike the sulfated polysaccharides of the red seaweeds, do not readily form gels except in the presence of borate and calcium ions. The polymers from Cladaphorales and Codiales appear to be related in general character but differing in fine detail. They may be regarded as sulfated xyloarabinogalactans in which ester sulfate groups can be present on D-galactose (C-3 or C-6) and arabinose (C-3). Polymers from the Ulvales,

Acrosiphoniales, and Ulitrichales may be regarded as sulfated glucuronoxylorhamnans in which sulfate can be present on L-rhamnose (C-2 and/or C-4) and D-xylose (C-2). In the unicellular green alga, *Acetabularia crenulata,* a glucuronoxylorhamnogalactan carries sulfate groups on D-galactose residues (C-4 or C-6), while the unicellular stage of the growth cycle of *Urospora worskioldii* contains, in addition to a sulfated glucuronoxylorhamnan, a sulfated mannan with sulfate groups at C-2 on the D-mannose units. Percival[29] has recently provided an excellent brief review of some of these Chlorophyceae polymers. Other general reviews that provide futher indications of the extraordinary complexity and diversity of the algal polysaccharide sulfates have been provided by Mackie and Preston,[30] McCandless and Craigie,[31] and Percival.[32]

b. Invertebrate Polysaccharide Sulfates

The presence of sulfated polysaccharides in invertebrate organisms, particularly in secretions of certain marine organisms, has been recognized for almost half a century. For the most part these compounds have attracted relatively little detailed attention and meaningful structures can rarely be assigned. The following examples will perhaps give some indication of the diversity of the polymers and probably does no more than skim the surface of a pool of compounds that almost certainly have an extremely wide natural distribution.

Polyglucan sulfates are known to be present in a number of marine mollusks. The gastropod, *Charonia lampas,* secretes a viscous sulfated mucus from the hypobranchial gland. The work of Egami's group in Japan[33] established that the sulfated polysaccharide, charonin sulfate, isolated from the mucus was actually a mixture of polyglucan sulfates. A "sulfur-rich" component has been partially characterized as a sulfated cellulose-like ($\beta 1 \rightarrow 4$-linked) polyglucan and indirect evidence from studies of its enzymic degradation and that of a synthetic cellulose polysulfate, indicated that the natural polymer possesses ester sulfate groups at positions 2 and 6 of the glucose residues.[34] The sulfur content of this material is variable, but accounts for up to three ester sulfate groups per disaccharide repeating unit. Much less is known about a second and "sulfur-poor" component, except that the linkage between the glucose units is $\alpha 1 \rightarrow 4$ and sulfur content accounts for one sulfate group per five or six glucose residues.

A somewhat similar mucus is secreted by the hypobronchial glands of the marine gastropod, *Buccinum undatum*. Infrared and enzymological and other studies suggested a cellulose-like structure containing sulfate groups equivalent to one group per glucose residue. Precise location of these groups is not established with certainty, but there is some support for the C-3 position. A summary account of this compound has been provided by its discoverer.[35]

Polyglucan sulfates are also present in the mantle mucus of the scallop, *Pecten maximus,* and in the odontophore of the whelk, *Busycon canaliculatum.*[35] In all the examples cited there are indications that the sulfated polymers may be associated covalently with protein.

Polyfucan sulfates are also a feature of some invertebrates, the best characterized of these being the thyonatan 4-sulfate present in the connective tissue of the sea cucumber, *Thyone briareus.*[36] The polysaccharide, which is covalently associated with protein, appears to be predominantly composed of $\alpha 1 \rightarrow 2$ linked L-fucopyran 4-sulfate residues. Polyfucan sulfates have also been isolated from the sea cucumber, *Stichopus japonicus* and from the jelly coats of the eggs of various sea urchins.[35,37,38]

Polygalactan sulfates have been identified in the egg jelly coat of at least one sea urchin,[37] *Echinus esculentus,* and in the tunicates, *Halocynthia roretzi* and *Styela plicata.*[39] Galactose appears to be in the L-form in all these examples (cf. seaweed polysaccharide sulfates) and there is some evidence to suggest that sulfate residues are located at C-4, although the tunicate polymers also appear to carry some sulfate at C-6 (or C-3).

A polyhexosamine sulfate has been demonstrated as a protein-linked polysaccharide component of the mucin of *Busycon canaliculatum*.[35,40] It appears to be a $\beta 1 \rightarrow 4$ linked polymer of alternating D-glucosamine and D-galactosamine residues. There are some indications that some, if not all, of the sulfate groups are present as *N*-sulfate (sulfamido) groups.

A number of other reports of invertebrate sulfated polysaccharides containing hexosamines as well as neutral sugars have been made.[35] For the most part information about structure is fragmentary, but the existence of such polymers serves to confirm the view expressed earlier about the wide natural distribution of invertebrate polysaccharide sulfates.

c. Sulfated Glycosaminoglycans

These polymers have been demonstrated[35,41] in members of the Porifera, Coelenterata, Platyhelminthes, Echinodermata, Tunicata, Mollusca, Annelida, Arachnida, Crustacea, Insecta, and Vertebrata, and it seems reasonably certain that they are a feature of most, if not all, members of the animal world. Detailed studies have generally been restricted to the mammalian polymers. These exist as proteoglycans in which sulfated glycosaminoglycan chains are covalently linked to a protein core and, in some cases, as giant proteoglycan aggregates in which other macromolecules also participate. The provision of a truly adequate account of these aggregates and their individual components would require a chapter in its own right and space limitations therefore allow mention only of those features of the glycosaminoglycan components that are essential to the current account. The reader is referred to an excellent recent review by Roden[42] for more detailed information.

Essentially, the sulfated glycosaminoglycans comprise chondroitin 4- and 6-sulfates, dermatan sulfate, keratan sulfate, heparin, and heparan sulfate. The chondroitin sulfates contain repeating disaccharide units composed of alternating D-glucuronic acid and *N*-acetyl-D-galactosamine residues, the latter being sulfated at C-4 (chondroitin 4-sulfate) or at C-6 (chondroitin 6-sulfate). The linkage between uronic acid and hexosamine is $\beta 1 \rightarrow 3$ and between hexosamine and uronic acid, $\beta 1 \rightarrow 4$.

Dermatan sulfate has a disaccharide repeating unit composed of alternating L-iduronic acid and *N*-acetyl-D-galactosamine 4-sulfate. However, some of the iduronic acid residues carry a sulfate group at C-2. Linkage between uronic acid and hexosamine is $\alpha 1 \rightarrow 3$, and between hexosamine and uronic acid, $\beta 1 \rightarrow 4$.

At least two different keratan sulfates are known and these differ in their tissue location and in the type of linkage by which they are bound to protein. The polysaccharide chains contain alternating residues of D-galactose and *N*-acetyl-D-glucosamine. Sulfate groups are present at C-6 of the hexosamine residues, while some of the galactose residues also carry sulfate in the C-6 position. The hexose to hexosamine linkage is $\beta 1 \rightarrow 4$, whereas the hexosamine to hexose link is $\beta 1 \rightarrow 3$.

Heparin and heparan sulfate are closely related to each other from the structural point of view. The basic repeating disaccharide unit of the former is composed of a uronic acid residue (either L-iduronic acid or D-glucuronic acid, with the former predominating) and D-glucosamine. Most of the hexosamine residues carry two sulfate groups, one located at C-6 and the other present as an *N*-sulfate (sulfamido) group at C-2. However, *N*-acetyl replaces *N*-sulfate in a small proportion of the hexosamine residues. Most residues are sulfated at C-2, but D-glucuronic acid residues generally remain unsulfated. Individual residues are joined by $1 \rightarrow 4$ linkages with those between iduronic acid and hexosamine being α and those between glucuronic acid and hexosamine being β.

Heparan sulfate consists of a family of molecules in which the polysaccharide chain has extensive regions where the D-glucosamine residues are acetylated and where the

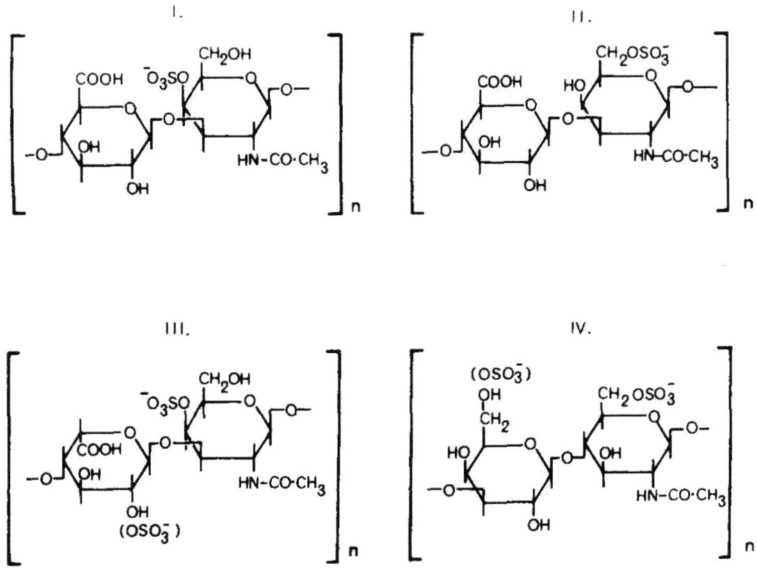

FIGURE 6. Major disaccharide repeating units present in I, chondroitin 4-sulfate; II, chondroitin 6-sulfate; III, dermatan sulfate; IV, keratan sulfate. See text for further details.

predominant uronic acid is glucuronic acid. In these regions, sulfate is present at C-6 of the hexosamines. Other regions occur in which *N*-sulfate groups and C-6 sulfate groups are present on the hexosamines and in which the predominant uronic acid, L-iduronic acid, is sometimes sulfated at C-2. Heterogeneity and polydispersity are common features of glycosaminoglycans, and this appears to be particularly true for heparan sulfate (and heparin). It is already known[43] that no clear subdivision exists between the two types of region. For example, regions have been noted[44] where *N*-sulfated and *N*-acetylated hexosamine residues alternate in the disaccharide units.

The principal repeating disaccharide units of these various sulfated glycosaminoglycans are shown in Figures 6 and 7. Meanwhile, it will be seen from this brief summary that this group of compounds has representatives in which ester sulfate groups are present at C-6 of *N*-acetyl-D-glucosamine, C-2 (*N*-sulfate) of D-glucosamine, C-4 and C-6 of *N*-acetyl-D-galactosamine, C-6 of D-galactose, C-2 of L-iduronic acid, and C-6 of D-glucosamine *N*-sulfate.

Although the preceding comments refer to glycosaminoglycans of mammalian origin, it seems reasonably certain that the analogous compounds detected in other representatives of the animal kingdom also occur in association with protein and possess basic structures that in general terms closely resemble those of the mammals. The amount of sulfur returned to soils and waters from connective tissue proteoglycans is incalculable, but must be enormous.

d. Other Sulfated Polysaccharides and Glycoproteins

Some reports of the presence of uronic acid-free sulfated polysaccharides in organisms other than invertebrates have been made. For example, a sulfated polygalactan-polyhexosamine, lorenzan sulfate, has been isolated from proteolytic digests of the jelly of the organs of Lorenzi of the elasmobranch fish *Squalus acanthias.*[45] The material contains roughly equal amounts of galactose and hexosamine (glucosamine and galactosamine, both probably acetylated), but the location of the sulfate groups (suf-

I.

II.

III.

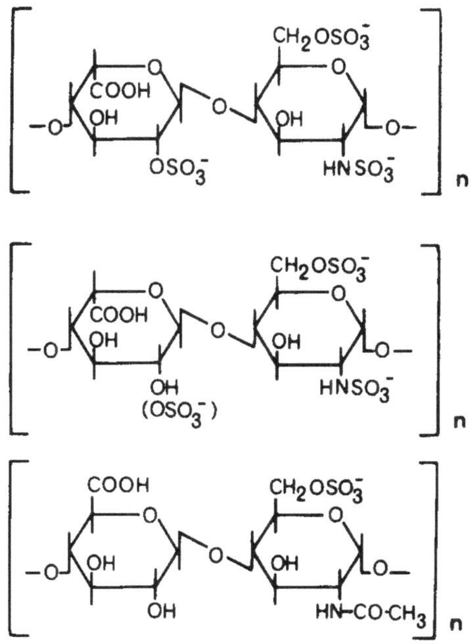

FIGURE 7. Major disaccharide repeating units present in heparin and heparan sulfates. I, Heparin; II and III, heparan sulfates. See text for further details.

ficient to account for one residue per monosaccharide residue) is not known. Because only trace quantities of uronic acid were present, the possibility of the material being a mixture of keratan sulfate and chondroitin sulfates was dismissed. In common with most of the other polysaccharide sulfates, it occurs naturally as a complex with proteins.

Hagfish *(Myxine glutinosa)* mucin contains protein-linked sulfated heteropolysaccharides with ester sulfate contents varying between 8 and 37%.[46] Galactose, fucose, mannose, glucosamine, and galactosamine comprise the monosaccharide units of these polymers.

Sulfated polysaccharides have been found in the cell wall of the halophilic bacteria, *Halobacterium halobium*[47] and *Halococcus morrhuae*.[48] Little appears to be known about the polymer from the former source, but that from the latter is very heterogenous and contains glucose, mannose, galactose, glucuronic, and galacturonic acids, *N*-acetyl-D-glucosamine, *N*-acetyl-D-galactosamine, gulosaminuronic acid, acetate, and glycine in addition to ester sulfate. The slime mold, *Physarum polycephalum,* also secretes a sulfated polysaccharide consisting mainly of D-galactose and traces of rhamnose.[49] The ester sulfate content could account for one sulfate group per two galactose residues. There is also a report[50] of the presence in *Clostridium welchii* NCTC 10578 of a capsular polysaccharide bearing some resemblance to dermatan sulfate.

These various compounds appear to be true macromolecular sulfated heteropolysaccharides, but much further structural work is needed before they can be clearly differentiated from another significant group of sulfated polymers, namely the sulfated glycoproteins. Here, in contrast to the proteoglycans described earlier, protein is the dominant compound to which relatively short oligosaccharide chains bearing ester sulfate groups are attached. The existence of such compounds has been known for many years, but the resolution of the problems associated with their isolation in pure form awaited the development of appropriate laboratory techniques. It was not until 1963 that Kent and Marsden[51] were able to isolate a sulfated glycoprotein from the mucin of the colon of the sheep. Since then the widespread distribution of such compounds has become increasingly obvious.

Yosizawa has written a useful review[52] on those sulfated glycoproteins studied prior to 1972, including examples separated from mammalian submaxillary glands, saliva, gastric juices, digestive tract, respiratory tract, cornea, and bile. Several of these have since been further studied, while new ones have been investigated from, for example, synaptosomes,[53] synaptosomal membranes,[54] myelin,[55] submandibular glands,[56] ovomucin,[57] uterus,[58] gastric tumors,[59] and embryonic liver and lung.[60]

Sulfated glycoproteins are also present in invertebrate organisms,[35] for example, in oyster viscera[61] and sea urchin embryos,[62] while structural and/or nonstructural sulfated viral glycoproteins are synthesized (coded by the viral genome) following the infection of host cells with herpes virus[63,64] or influenza virus.[65,66]

The functions of the sulfate groups in the sulfated glycoproteins are far from clear. Kent and his colleagues[67] have attempted to argue that they may serve in an informational capacity in the regulation of the biosynthesis and turnover of glycoproteins and hence (presumably) in the regulation of development and differentiation. Some support for this viewpoint comes from the finding[68] that the large, external transformation-sensitive glycoprotein (LETS protein, fibronectin) is sulfated in normal hamster fibroblast cells in culture and in the culture medium, but is not so sulfated in the transformed cell. Meanwhile, Kent et al.[67] have pointed out that the possession of suitable sulfatase enzymes might be of great help in furthering our present inadequate understanding of the importance of the sulfate groups to these compounds.

At the present time, the structure of any individual sulfated glycoprotein is not known with certainty. However, ester sulfate groups appear generally to be located at the 6-position of either *N*-acetyl-D-glucosamine of *N*-acetyl-D-galactosamine in those cases where experimental evidence has been obtained. For example, one of the principal sulfated oligosaccharide chains (oligosaccharide II) of ovomucin consists of a trisaccharide unit, *N*-acetylneuraminyl (2→3)-galactosyl(1→3)-*N*-acetyl-D-galactosamine 6-sulfate, the sulfated hexosamine being linked to protein via serine or threonine.[57] Additional information about sulfated glycoproteins can be obtained from a recent review by Marshall.[69]

II. GLYCOSULFATASE

The Enzyme Nomenclature Committee list describes this enzyme as a sugar-sulfate hydrolase hydrolysing D-glucose 6-sulfate, but acting also on other sulfates of mono- and disaccharides and on adenosine 5′-sulfate. The description, although basically correct, is really of very little help in clarifying the confused situation that exists in relation to these and certain other enzymes participating in the desulfation of sulfated carbohydrate residues. For example, is an enzyme that removes a sulfate group from one of the two D-galactose residues of the tetrasaccharide, neocarratetraose disulfate, a glycosulfatase or is such a classification inadequate? It will become apparent that much remains to be done before the true specificities and natural substrates of these enzymes allow a truly adequate nomenclature.

Originally termed "glucosulfatase" by its discoverers,[70] the enzyme was later renamed glycosulfatase on the basis that sulfate esters of monosaccharides other than glucose could also serve as substrates.[71-73] Work from the Cardiff laboratories, together with that from Egami's laboratory and from others,[74] suggested that glycosulfatases from various marine mollusks were active toward the 6-sulfate esters of D-galactose, L-fucose, *N*-acetyl-D-glucosamine and *N*-acetyl-D-galactosamine, as well as D-glucose 3- and 6-sulfates and adenosine 5′-sulfate. However, these early conclusions were based on work with fairly crude enzyme preparations with the possibility that more than one carbohydrate sulfatase was present. The later availability of a highly purified preparation of an enzyme from the digestive gland of *Charonia lampas* added weight to the possibility because, although D-glucose 3- and 6-sulfates and *N*-acetyl-D-glucosamine 6-sulfate were substrates for the enzyme, D-galactose 4- and 6-sulfates, *N*-acetyl-D-galactosamine 6-sulfate and L-fucose monosulfate (mainly the 3-sulfate) were not.[75] It is interesting that all the molluscan enzymes exhibit relatively high K_m values toward the various substrates used and one is tempted to wonder if the true substrates still await discovery.

Bearing in mind the unresolved specificity problems of molluscan glycosulfatases, particular care is needed in defining the specificity of the enzymes from bacteria, particularly as impure enzyme preparations have generally been studied. The position is further complicated by the possibility that some bacterial glycosulfatases may have wider roles to play in the biodegradation of polymeric materials containing sulfated monosaccharide units and this point will be elaborated later in relation to the degradation of two such polymers, carrageenan and keratan sulfate. Indeed, the whole field of microbial glycosulfatases now cries out for reinvestigation and reappraisal and it is hoped that this present brief account might provide the necessary stimulus for this.

The first account[76] of a microbial glycosulfatase was provided in 1932 by Tanko who found that extracts of cells of *Bacterium fluorescens non liquifaciens* (syn. *Pseudomonas non liquifaciens)* that had been grown on a medium containing "chondroitin sulfate", hydrolysed glucose, and sucrose monosulfates (presumably 6-sulfates). It is worth noting that sucrose could be hydrolysed by the cell extracts so that the possibility could not be excluded that the sucrose monosulfate preparation was first hydrolysed to give glucose monosulfate before the sulfatase could act.

Interest in this bacterial enzyme was short lived and almost 20 years were to elapse before Yamashina[77] investigated the possibility that other microorganisms could produce a similar enzyme. *Aspergillus oryzae, Pseudomonas fluorescens, P. aeruginosa, Escherichia coli* and yeast failed to produce such an enzyme under the growth conditions employed, but a mold, identified as *Trichoderma viride,* grew well on a synthetic medium containing D-glucose 6-sulfate and centrifuged cell extracts hydrolysed the ester.

Again interest waned until some observations made in the Cardiff laboratories led to further work. Briefly, during studies on the metabolic fate in rats of ^{35}S-labeled samples of *N*-acetyl-D-glucosamine 6-sulfate[78] and *N*-acetyl-D-galactosamine 6-sulfate,[79] it was noted that rat feces, homogenized and incubated in acetate buffer containing the esters, liberated appreciable quantities of $^{35}SO_4^{2-}$ ions from them. Little or no release of sulfate was observed if the experimental animals had been treated with oral doses of sulfasuxidine and terramycin over the previous 70-hr period. Subsequently, it was shown[80] that rat fecal suspensions could also remove the sulfate group from the 6-sulfate esters of D-glucose and D-galactose as well as from ^{35}S-labeled cerebroside 3-sulfate (galactose 3-sulfate residue).[81] Mixed cultures of the fecal flora grown anaerobically at 37° C in nutrient broth and transferred to defined media containing either of the two monosaccharide sulfates, hydrolysed those esters only after a lag period varying between 24 and 48 hr. In contrast, hydrolysis began immediately if the organisms had previously been grown on broth containing D-glucose 6-sulfate. A Gram-positive, microaerophilic rod, tentatively identified as a *Lactobacillus* species, was finally isolated which was responsible for the sulfatase activity.[80] However, the organism had complex growth requirements and was not considered suitable as a potential source of carbohydrate sulfatases.

Attention now turned to the earlier work of Yamashina[77] on *Trichoderma viride* and a study of the ability of the mold to produce glycosulfatase was undertaken.[82] When grown on defined media containing D-glucose or D-galactose as carbon source and K_2SO_4 as sulfur source or with glucose and methionine as carbon/sulfur source, no glycosulfatase activity could be detected toward the ^{35}S-labeled 6-sulfates of D-glucose or D-galactose. However, growth on defined media containing either of the esters (unlabeled) led to the production of a glycosulfatase that could hydrolyse both. The lag period associated with the induction of the enzyme was very variable and a particular culture sequence was necessary before the lag period became reasonably constant. It was concluded that the variation reflected the different levels of reserve stores of sulfur in the microorganisms and supporting this possibility was the observation that the mold readily produced choline sulfate, known to serve as an important store of sulfur in many fungi (see Volume 1, Chapter 2, Section III. A.).

Cell extracts were employed to examine some properties of the glycosulfatase, the activity of which was measured by estimating liberated SO_4^{2-} ions turbidimetrically[83] or $^{35}SO_4^{2-}$ by the electrophoresis/scanning method,[82,84] or the release of D-galactose or D-glucose by the galactose oxidase or glucose oxidase methods.[82] Desulfated hexose and SO_4^{2-} ions were released in equimolar amounts from the substrates, thus establishing that a simple hydrolase was involved. The enzyme was maximally active at 28° C, pH 7.4 to 7.9, and at a substrate concentration of about 23 m*M*. Although K_m values were not determined it is clear that the value for D-glucose 6-sulfate would be considerably less than those obtained for molluscan glycosulfatases.[75] *Trichoderma* glycosulfatase was devoid of activity toward D-glucose 3-sulfate, a further point of contrast to the molluscan enzymes.

No further studies on the *Trichoderma* enzyme have been made. Moreover, with one exception (see Section III), there appear to have been no further attempts to search for a microbial glycosulfatase specifically concerned with the hydrolysis of D-glucose 6-sulfate. However, a brief mention will be made here of two microbial enzyme systems that are very relevant to the present topic but which were discovered through attempts to find enzymes that could participate, on the one hand, in the biodegradation of the seaweed polysaccharide carrageenan and, on the other, in the biodegradation of keratan sulfate. It will be recalled from earlier sections (I.A. 5.a and c) of this chapter that both biopolymers contain sulfated galactose units.

In an attempt to employ definitive degradative enzyme systems to unravel the complexity of carrageenan structure, Weigl and Yaphe[85] purified an enzyme (from the culture medium) from *P. carrageenovora* that degraded κ-carrageenan to a homologous series of oligosaccharides, the major product being the disaccharide, neocarrabiose 4-sulfate (3,6-anhydro-α-D-galactose {α1→3}-D-galactose 4-sulfate, Figure 8). Subsequently, cell extracts of the bacterium were shown[86] to possess a sulfatase enzyme (termed a glycosulfatase by the investigators) that would remove sulfate from the 4-position of the D-galactose residue of the disaccharide. Surprisingly, the claim was made that the enzyme also hydrolysed D-galactose 6-sulfate, but not the corresponding 4-sulfate. Recently the sulfatase has been considerably purified and separated from the coexisting κ-carrageenase.[87] The purified preparation not only removed the sulfate group from neocarrabiose 4-sulfate, but also one of the two sulfate groups present as the corresponding tetrasaccharide, neocarratetraose disulfate. Unfortunately, no attempt was made to test the ability of the enzyme to hydrolyse monosaccharide sulfate esters, so that the claim[86] that D-galactose 6-sulfate was a substrate for the enzyme has not been substantiated. Further work is needed; meanwhile a more detailed account of the *P. carrageenovora* sulfatase is provided in Section IV.

A similar unresolved situation is associated with the participation of sulfatase enzymes in the microbial biodegradation of keratan sulfate. Studies[88,89] on the soil bacterium *Actinobacillus* IFO-11310 grown in liquid medium containing shark keratan sulfate (most of the galactose residues sulfated at C-6), showed that three distinct sul-

I.

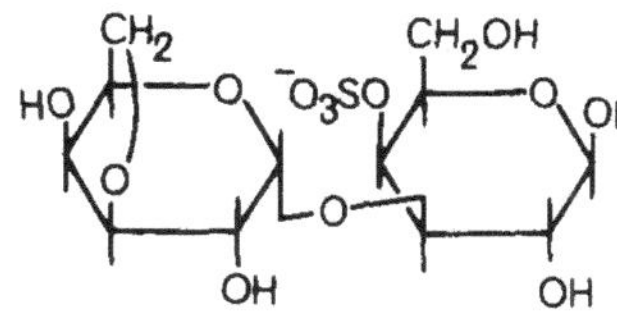

II.

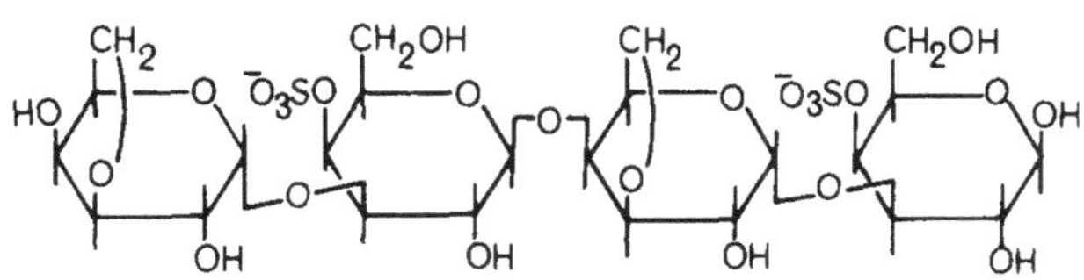

FIGURE 8. Di- and tetrasaccharides isolated following the action of the carrageenase of *Pseudomonas carrageenovora* on κ-carrageenan. I, neocarrabiose 4-sulfate; II, neocarratetraose disulfate. See text for further details.

fatases were present in cell extracts. One of these was able to remove the ester sulfate group from the hexosamine moiety of the disaccharide, *N*-acetyl-D-glucosamine 6-sulfate (β1→3)-D-galactose and from the nonreducing *N*-acetyl-D-glucosamine 6-sulfate residue of the tetrasaccharide, *N*-acetyl-D-glucosamine 6-sulfate (β1→3)-D-galactose 6-sulfate (β1→4)-*N*-acetyl-D-glucosamine 6-sulfate (β1→3)-D-galactose (Figure 9). This enzyme was also active toward the monosaccharide, *N*-acetyl-D-glucosamine 6-sulfate.

A second sulfatase removed the sulfate group from the D-galactose 6-sulfate residue of the trisulfated tetrasaccharide shown in Figure 9, and was also active toward *N*-acetyl-D-galactosamine 6-sulfate and (to a lesser extent) D-galactose 6-sulfate. Finally, a third sulfatase, active toward D-glucose 6-sulfate, was also present in cell extracts but has not been studied further. More detailed information about these sulfatases and their role in the biodegradation of keratan sulfate is provided in Section VI.

The question now arises as to whether any of these sulfatases from *P. carrageenovora* and *Actinobacillus* IFO-11310 can be regarded as glycosulfatases analogous to those present in mollusca or in *T. viride*. At the present time insufficient evidence is available to enable an unequivocal answer to the question and there is an urgent need for detailed specificity and kinetic studies aimed at defining the true natural substrates for all the enzymes mentioned in this section. It will be recalled from Section I.A. of this chapter that sulfate groups are to be found on a large number of different monosaccharide entities, including D- and L-hexoses, anhydrohexoses, pentoses, and hexosamines. How many of these sulfated entities can serve as substrates for "glycosulfatase"? Possibly a whole series of such enzymes of differing specificities is awaiting discovery. Furthermore, sulfated monosaccharides do not appear to occur naturally as such, but rather as components of larger molecules ranging in complexity from simple sulfated glycosides such as *p*-coumaryl-D-glucose 6-sulfate (Figure 2) to complex heteropolysaccharides such as carrageenan and keratan sulfate. The question is therefore raised as to whether sulfated monosaccharide residues are first released by the action of appropriate glycosidases before undergoing desulfation or, alternatively, are desulfated while present as a component of a larger, intact, or partly degraded molecule.

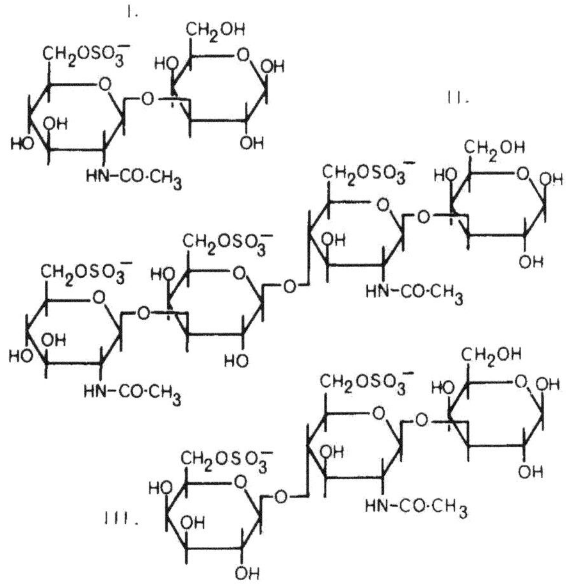

FIGURE 9. Di-(I), and tetra-(II) saccharides isolated following the action of the endo-β-galactosidase of *Escherichia freundii* on various keratan sulfates. Trisaccharide (III) prepared by sequential degradation of keratan sulfate with the endo-β-galactosidase, *N*-acetyl-D-glucosamine 6-sulfatase, and exo-β-*N*-acetylglucosaminidase of *Actinobacillus* IFO-13310. See text for further details.

III. GLUCONATE 6-SULFATASE AND GLYCERATE 3-SULFATASE

For reasons that will become obvious, it is convenient to deal with these two enzymes in the same section. Both were discovered as a result of further searches for microbial glycosulfatases, during the course of which it was noted that *P. fluorescens* N.C.I.B. 8248 was able to utilize D-glucose 6-sulfate as sole carbon and sulfur source for growth. A technique was subsequently developed that enabled cells to become adapted to growth on a basal salts medium containing the ester so that a constant lag period could be obtained. The adapted strain grew well at 37° C whereas the unadapted strain showed poor growth at that temperature.[90]

Optimum conditions for growth and utilization of D-glucose 6-sulfate (pH, temperature, D-glucose 6-sulfate concentration, culture age) were established and the bacterium was then grown in the presence of 60 m*M* D-glucose 6-(^{35}S)sulfate. At the termination of growth some 13% of the radioactivity had disappeared from the medium, 50% of the starting material was still present and the remainder was made up of three labeled metabolites that could be separated by paper electrophoresis. One of the metabolites accounted for 1% only of the total radioactivity of the medium and was not studied further. A second metabolite (4% of the radioactivity in the medium) was identified as D-gluconate 6-(^{35}S) sulfate, while the third one (32% of the radioactivity in the medium) was identified as D-glycerate 3-(^{35}S) sulfate.

Subsequently, extracts were prepared from adapted cells taken at different time intervals during growth on D-glucose 6-sulfate and were tested for their ability to desulfate D-gluconate 6-(^{35}S) sulfate, DL-glycerate 3-(^{35}S) sulfate, and D-glucose 6-(^{35}S) sulfate. The first two esters were hydrolysed by extracts, with maximum activity occurring

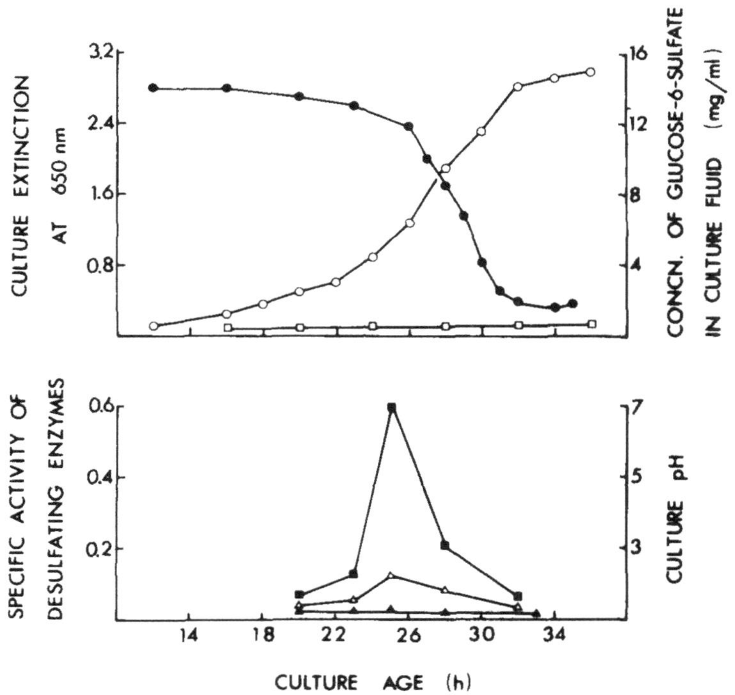

FIGURE 10. Relationship between culture extinction and utilization of D-glucose 6-sulfate during growth of *Pseudomonas fluorescens* on 1.6% D-glucose 6-sulfate. ○, Culture extinction; ●, utilization of D-glucose 6-sulfate; □, culture extinction in the absence of D-glucose 6-sulfate; ■, ability of cell extracts to desulfate DL-glycerate 3-(^{35}S)sulfate; △, ability of cell extracts to desulfate D-gluconate 6-(^{35}S)sulfate; ▲, ability of cell extracts to desulfate D-glucose 6-(^{35}S)sulfate. See text for further details.

with those taken shortly after the commencement of exponential growth and concomitant utilization of D-glucose 6-sulfate (Figure 10). D-Glucose 6-(^{35}S) sulfate was not hydrolysed by any of the cell extracts.

Maximum sulfatase activity toward DL-glycerate 3-(^{35}S) sulfate (1 m*M*) was obtained at pH 7.0 in glycylglycine-NaOH buffer. Inhibition by excess substrate was very marked at concentration greater than 1 m*M*. Although K_m values were not calculated the enzyme clearly has a relatively high affinity for the substrate. When cells were grown in basal salts medium containing D-glucose and K_2SO_4 as the respective sources of carbon and sulfur, relatively low enzyme activity toward D-gluconate 6-sulfate (10% of that of adapted cells) could still be detected, but no desulfation of DL-glycerate 3-(^{35}S) sulfate or D-glucose 6-(^{35}S) sulfate occurred. This suggests that separate enzymes are responsible for the hydrolysis of the glycerate and gluconate sulfate esters.

In a further study[91] *P. fluorescens* N.C.I.B. 8248 was grown in basal salts medium containing glycerol as carbon source and D-glucose 6-(^{35}S) sulfate as sole sulfur source. At the termination of growth some 12% of the initial total radioactivity had disappeared from the medium, 57% was still present as starting material while the remainder consisted of a single metabolite identified as D-gluconate 6-(^{35}S) sulfate. Separate experiments established that cell extracts were able to oxidize D-glucose 6-(^{35}S) sulfate to the corresponding gluconate derivative. Earlier work[92] had established that D-glucose 6-sulfate can be oxidized to D-gluconate 6-sulfate by purified yeast D-glucose 6-phosphate dehydrogenase, although at a relatively low rate. However, attempts to establish that this was the type of enzyme operating in the case of *P. fluorescens* were unsuccessful. On the other hand some evidence was obtained to support the possibility that the responsible enzyme was similar to the glucose dehydrogenase known to be present in some pseudomonads.[93]

Other experiments showed that cell extracts of *P. fluorescens* were able to liberate $^{35}SO_4^{2-}$ ions from D-gluconate 6-(^{35}S) sulfate, but not from D-glucose 6-(^{35}S) sulfate. Maximum desulfation of the former was observed in cells harvested just after the onset of exponential growth and utilization of D-glucose 6-sulfate as sulfur source for growth. These extracts were devoid of sulfatase activity toward DL-glycerate 3-(^{35}S) sulfate, confirming that two distinct enzymes are involved in the desulfation of gluconate and glycerate sulfate esters.

Collectively, these various results indicate that this bacterium utilizes D-glucose 6-sulfate as sole source of carbon and sulfur for growth by a sequence of reactions identical with or analogous to those operating in the metabolism of D-glucose 6-phosphate by the Entner-Doudoroff pathway. With the sulfate group of D-glucose 6-sulfate in a position remote from events occurring during the first part of the pathway the reactions proceed as far as the production of two 3-carbon fragments, one of which (glycerate 3-sulfate) must lose its ester sulfate group before further metabolism can occur. Sulfur for growth can be obtained at the D-gluconate 6-sulfate or D-glycerate 3-sulfate stage of the degradation. In contrast, when D-glucose 6-sulfate is used to supply only sulfur for the growth of the bacterium, that sulfur comes from the hydrolysis of the first oxidation product, D-gluconate 6-sulfate. Under such growth conditions, little or no further metabolism of the gluconate sulfate occurs and, in support of this contention, the ability of cell extracts of the bacterium to metabolize that compound (other than by hydrolysis) is very low in comparison with extracts of cells grown on D-glucose 6-sulfate as sole source of both carbon and sulfur.[91]

No further studies on the two sulfatases have been made. Nevertheless, the two enzymes are of considerable interest in that they open up new possibilities in relation to the biodegradation of sulfated D-glucose moieties. Moreover, the glycerate sulfatase may possibly turn out to have some significance in relation to the biodegradation of the sulfated glycerol residues of the sulfolipids from the halophilic bacteria, *Halobacterium cutirubrum*[12] and *H. marismortui.*[94] Clearly neither of the enzymes can be classified as glycosulfatases.

IV. κ-CARRAGEENAN 4-SULFATASE

A number of microbial sulfatases capable of participating in the biodegradation of algal polysaccharide sulfates must surely exist but, at the present time, only one such enzyme has been described. The structural complexity and diversity of algal polysaccharide sulfates (see Section I.A.5.a) is such that enzyme systems of defined specificity could serve as powerful tools in the elucidation of the fine structures of the polymers. It is not absolutely clear, therefore, why little interest has been taken in this line of approach, particularly as Araki and his colleagues, for example, made such a promising start on the use of bacterial enzymes in the study of the nonsulfated polymer, agarose.[95]

Yaphe and his colleagues led the quest for microbial sources of enzymes that could degrade sulfated seaweed polysaccharides.[96] An enrichment culture technique was employed to select organisms that were capable of degrading carrageenans prepared from a variety of red seaweeds. Finally, a pure culture of a Gram-negative rod was obtained which was able to hydrolyse the κ-component of the carrageenan isolated from the seaweed, *Chondrus crispus.* The microorganism was subsequently designated[85] as *P. carrageenovora* and was shown to produce at least two glycosidases (carrageenases),

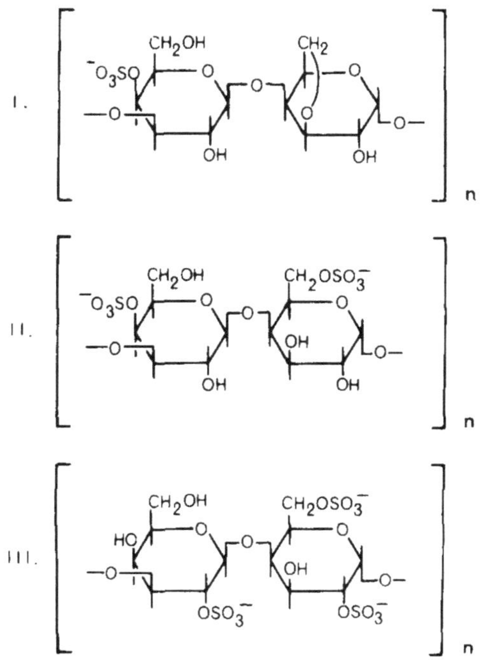

FIGURE 11. Major disaccharide repeating units present in I, κ-carrageenan; II, μ-carrageenan; III, λ-carrageenan. See text and Reference 97 for further details.

one of which was specific for κ-carrageenan and the other for "λ-carrageenan." A further word of explanation concerning these carrageenan fractions might be useful at this point.

On the basis of their idealized chemical structures, Rees[97] has considered the carrageenan of *C. crispus* as consisting of three components. All possess a repeating structure of alternate 3- and 4-linked galactose units which is modified by sulfation or anhydride formation. Thus the idealized structure of κ-carrageenan is envisaged as containing 3-linked β-D-galactose 4-sulfate and 4-linked 3,6-anhydro-D-galactose. This carrageenan fraction is precipitated from solution by KCl. μ-Carrageenan and λ-carrageenan are not so precipitated. The μ-fraction contains 3-linked β-D-galactose 4-sulfate and 4-linked α-D-galactose 6-sulfate, and κ-carrageenan is derived from this as a result of the removal of the 6-sulfate residues by an eliminase enzyme (yielding 3,6-anhydro-D-galactose) that is present in the seaweed. λ-Carrageenan mainly consists of 3-linked β-D-galactose 2-sulfate and 4-linked α-D-galactose 2,6-disulfate (see Figure 11 for structures). This is a somewhat oversimplified structural picture of the components present when carrageenan is isolated from a particular seaweed. Carrageenans from other sources do not necessarily follow this pattern precisely (see Section I.A.5.a). Meanwhile, it should be noted that the term "λ-carrageenan", as used in the reports of the studies of Yaphe's group, actually refers to a mixture of λ- and μ-forms.

Detailed studies[85] of the carrageenase activity of *P. carrageenovora* involved the growth of the bacterium (shake culture at 25° C) in a basal salts medium containing 0.25% commercial carrageenan (all fractions presumably present) which served as carbon and sulfur source. Maximum carrageenase activity toward κ-carrageenan was obtained during the stationary phase of growth, whereas production of λ-carrageenase followed the growth curve. The κ-carrageenase was purified some 30-fold from the culture medium and attempts were made to isolate the product of its action toward κ-carrageenan. An homologous series of four oligosaccharides as well as an enzyme resistant component were separated from the enzyme digests, and the series was shown to consist of the sulfated disaccharide, neocarrabiose 4-sulfate, together with the corresponding di-, tri-, and tetrasulfated tetra-, hexa-, and octasaccharides. No such detailed study was made of the λ-carrageenase produced by the bacterium, but the enzyme appeared to cause a rapid drop in the viscosity of solutions of the polymer with only a relatively low production of reducing power. Collectively, these results enabled the authors to make some useful comments on the structure of the κ-carrageenan fraction.

In a later study[86] it was shown that the cells of *P. carrageenovora* (grown on the usual carrageenan-containing medium) contained a sulfatase enzyme. Sonication of cell suspensions released the enzyme, which was then partially purified by removing nucleic acid and fractionating the nucleic acid-free extract with ammonium sulfate. Some reference to this enzyme has already been made (Section II) and it will be recalled that it was able to remove the 4-sulfate group from the D-galactose residue of neocarrabiose 4-sulfate. D-Galactose 6-sulfate, but not the corresponding 4-sulfate, could be hydrolysed by the preparation, an unconfirmed finding so odd that one is tempted to conclude that the preparation must have contained a second sulfatase (a glycosulfatase?) exhibiting activity toward the 6-sulfate ester of D-galactose (and possibly other monosaccharides).

Later work by McLean and Williamson[87] resulted in the preparation of the sulfatase enzyme in a highly purified form and it was confirmed that neocarrabiose 4-sulfate was a substrate. However, the enzyme could also remove one of the two sulfate groups present in the corresponding tetrasaccharide, neocarratetraose disulfate (Figure 8), and it was established that the susceptible group was the one present at C-4 of the D-galactose residue proximal to the nonreducing end of the tetrasaccharide. Interestingly, the enzyme was completely inactive in the absence of NaCl. Maximum activity toward neocarrabiose 4-sulfate was obtained at a concentration of 0.25 *M* NaCl and pH 7.0, and the K_m value obtained for this substrate was 0.21 m*M*. Enzyme activity was strongly inhibited by phosphate ions. From the information provided by the investigators it is not possible to deduce which of the two substrates (di- or tetrasaccharide) is the better. However, the sulfated tetrasaccharide is certainly a most effective substrate.

It was particularly unfortunate that McLean and Williamson did not take an opportunity to examine the activity of the enzyme toward the 4- and 6-sulfate esters of D-galactose, as any coexisting "glycosulfatase" specific for monosaccharide sulfates might have been expected to have been eliminated during the purification of the neocarrabiose sulfatase.

In a parallel study[98] the same two investigators purified the κ-carrageenase present in the cell-free medium of cultures of *P. carrageenovora* and confirmed that the enzyme was not active toward the λ-component of *C. crispus* carrageenan. Indeed, in contrast to Weigl and Yaphe,[85] they failed to find any depolymerizing activity toward this substrate in the bacterial culture medium. It was suggested that this failure might reflect slight differences in the method of culture or some genetic change in the strain of organism supplied to them. However, other workers[99] have confirmed the existence of "λ-carrageenase" activity in the bacterium and showed that two distinct glycosidase-active fractions could be separated from the culture fluid. One of these (fraction I) produced a rapid decrease in the viscosity of "λ-carrageenan" solutions but very little increase in reducing power, suggesting that some type of endoglycosidase was involved. The other (fraction II) caused a significant increase in reducing power of the substrate solution as well as a decrease in viscosity. This fraction could be further subdivided into two active subfractions, only one of which exhibited any significant ability to liberate reducing materials. It was tentatively concluded that this particular subfraction contained an oligosaccharase acting on polymer segments produced by the other subfraction.

A somewhat unsatisfactory situation now remains to be resolved. In the case of the κ-component of the carrageenan isolated from *C. crispus*, it seems clear that a specific carrageenase is released into the culture medium that enables the greater part of the polysaccharide to be degraded to sulfated oligosaccharides. The smallest of these, neocarrabiose 4-sulfate, presumably enters the bacterial cell where it undergoes desulfation to yield neocarrabiose, thereby providing both sulfur and carbon for growth. The precise significance of the ability of the purified sulfatase to hydrolyse one of the two ester sulfate groups of the tetrasaccharide, neocarratetraose disulfate, remains unclear. Can this tetrasaccharide also be taken into the bacterial cell for desulfation, and if so how is the monosulfated product of the sulfatase action metabolized further? The work of McLean and Williamson[87] indicates that the conversion of neocarratetraose disulfate to neocarrabiose 4-sulfate by κ-carrageenase is a slow process, whereas the desulfation of the tetrasaccharide to neocarratetraose monosulfate is rapid. The investigators suggest that neocarratetraose disulfate may also be readily taken up by the bacterial cells and the β1→4 linkage between D-galactose and 3,6-anhydro-D-galactose might be hydrolysed by a β-galactosidase. The cells do indeed contain a β-galactosidase active toward *p*-nitrophenyl-β-D-galactoside, but it is not known whether it is active toward either neocarratetraose di- or monosulfates. If not active toward the disulfated tetrasaccharide the sequence of events would need to be sulfatase: glycosidase:sulfatase. If active toward that tetrasaccharide the sequence would be glycosidase:sulfatase. In either case the total desulfation of the tetrasaccharide need involve only the one sulfatase enzyme. Of course, it is also not known whether sulfated hexa- or octasaccharide can serve as substrates and much work remains to be done before a clear picture emerges with respect to the biodegradation of this particular κ-carrageenan.

The possible role of sulfatases in the biodegradation of the soluble λ- and μ-components of *C. crispus* carrageenan also presents intriguing problems. Depolymerizing glycosidases different from those attacking the κ-component are obviously involved and the question arises as to whether the κ-carrageenan 4-sulfatase involved in the biodegradation of the κ-component is able to remove sulfate from the sulfated fragments produced during λ- and μ-carrageenan biodegradation. It will be recalled that such fragments may contain ester sulfate groups at C-4, C-6, and C-2 of D-galactose as well as some 2,6-disulfated D-galactose residues. In the light of work on the bacterial degradation of animal sulfated glycosaminoglycans to be described in later sections of this chapter, it would perhaps by surprising if the carrageenan sulfatase did indeed exhibit enzyme specificity sufficiently broad to encompass all the positional dispositions of ester sulfate possible in carrageenan. Perhaps the finding in *P. carrageenovora* of sulfatase activity towards D-galactose 6-sulfate[86] is providing a pointer to the existence of other sulfatases that participate in the degradation of the λ- and/or μ-components of this interesting polydisperse sulfated polysaccharide.

The wide natural distribution and great structural diversity of algal polysaccharide sulfates has been pointed out in an earlier section of this chapter (Section I.A.5.a). Among these polymers, examples can be found in which sulfate is present at positions 2,3,4, and 6 of D-galactose at C-2 in 3,6-anhydro-D-galactose, D-mannose, L-rhamnose, and D-xylose; at C-3 of L-fucose and D-arabinose; C-4 of L-fucose; and at C-4 and C-6 of L-rhamnose. The task of unraveling the ways in which microbial sulfatases participate in the biodegradation of such a diverse group of complex polymers seems truly daunting. However, it will have to be undertaken sooner or later if a better understanding of the contribution made by polymers and microorganisms to the recycling of sulfur is to be reached.

V. CHONDRO-4-SULFATASE AND CHONDRO-6-SULFATASE

These two enzymes are involved in the biodegradation of the sulfated glycosaminoglycan chains of connective tissue proteoglycans. Chondro-4-sulfatase participates in the degradation of chondroitin 4-sulfate and dermatan sulfate chains, whereas the 6-sulfatase is concerned in the degradation of chondroitin 6-sulfate chains.

A brief and somewhat oversimplified account of these glycosaminoglycans has already been presented (Section I.A.5.c and Figures 6 and 7) and little more needs to be added here except to remind the reader that early workers in the field were not aware that three isomeric forms of a basic polymer existed. The terms chondroitin sulfate and chondrosulfatase that are common to the older literature reflects this fact and will sometimes be used in the present account.

A. Discovery of the Enzymes

Bacterial sulfatase activity toward chondroitin sulfate was noted as long ago as 1914 when Neuberg and Rubin[100] showed that a mixed culture of "putrefying bacteria" could liberate sulfate from the material. Much later work from the same laboratory[101] established that a similar desulfating activity was present in the cells of pure cultures of *Bacillus fluorescens non liquifaciens* (syn. *P. fluorescens), Bacterium pyocyaneum* (syn. *P. pyocyanea),* and *Bacillus proteus* (syn. *Proteus vulgaris*). Desulfation was accompanied by depolymerization of the substrate and release of reducing materials, and all these activities were still present in acetone-dried preparations of cell extracts. The name chondrosulfatase was assigned to the sulfatase enzyme and the term chondroitinase later became generally accepted for the system responsible for the depolymerizing activity.

Following these early observations a number of other workers noted the presence of similar enzymes in microorganisms, including strains of *Micrococcus pyogenes aureus* (syn. *Staphylococcus aureus),*[102,103] *Clostridium welchii,*[104] and the mold, *Penicillium spinulosum.*[105] Much of the interest in the enzymes stemmed from a hypothesis[106] that related dental caries to the presence of microorganisms capable of degrading the "mucopolysaccharide" component (i.e., sulfated glycosaminoglycan) of dentine, and numerous attempts were made to substantiate the hypothesis.[102-110] By and large very little of substance emerged from these various studies and it will later be apparent that some were even somewhat misleading, at least in substance if not in intention.

B. Sulfatase Activity in *Proteus vulgaris*

Systematic study of the sulfatase activity toward chondroitin sulfate isomers first began in the Cardiff laboratories. A preparation of ox tracheal cartilage chondroitin sulfate (consisting mainly of chondroitin 4-sulfate) was employed as substrate in a search for sulfatase activity in defined bacteria.[111] Cells were grown at 25° C in nutrient broth containing 1% of Oxoid® peptone, 0.3% Oxoid® Lab Lemco, and 0.1% NaCl, and cell extracts were then tested for ability to degrade the chrondroitin 4-sulfate preparation. No attempt was made to include substrate in the growth medium. Four strains of *P. vulgaris* obtained from the U.K. National Collection of Type Cultures were able to liberate sulfate and reducing materials from the substrate, whereas two strains of *Pseudomonas fluorescens,* two strains of *P. ovalis,* and a strain of *Alcaligenes metalcaligenes* were not.

The most active of the *Proteus vulgaris* strains (NCTC 4636) was selected for further study and some general properties of the sulfatase component of some partially purified preparations were examined. Sulfatase activity was maximal at pH 7.0 and a concentration of 0.5% substrate. Heparin, several seaweed polysaccharide sulfates, hexose and hexosamine sulfates, and aryl-, alkyl-, and steroid sulfates were not hydrolysed by the enzyme preparation.

Subsequently strain NCTC 4636 was grown on a large scale (again in the absence of chondroitin sulfate) and the interdependence of the sulfatase and chondroitinase enzyme was determined.[112] Earlier studies[113] had provided a useful clue to this interdependence by showing that resting cells of a different strain of *P. vulgaris* were unable to desulfate chondroitin sulfate in the presence of sulfite or hydroxylamine, but could still depolymerize the substrate. Acting on this clue, Dodgson and Lloyd[112] separated the sulfatase activity of strain 4636 from the associated chondroitinase and demonstrated that the former was completely inactive toward polymeric chondroitin 4-sulfate in the absence of the latter. Furthermore the sulfatase was active toward chondroitin 4-sulfate that had been exhaustively degraded to sulfated oligosaccharides (mainly tetrasaccharide) by testicular hyaluronidase. Phosphate inhibited the sulfatase activity of preparations that also contained chondroitinase, but did not affect the activity of the latter enzyme. Collectively, these results firmly established that depolymerization was a prerequisite for removal of sulfate groups.

Evidence from other studies[114,115] (one of them on a different strain of *P. vulgaris*) suggested that depolymerization of chondroitin 4-sulfate by chondroitinase involved a simple hydrolytic attack. However, it was later established[116,117] (and subsequently confirmed[118]), using preparations from *P. vulgaris* cells that had been grown in the presence of chondroitin sulfate, that depolymerization of chondroitin 4- and 6-sulfates proceeded via an elimination type of mechanism with the major products being, respectively, the sulfated Δ4,5-unsaturated disaccharide, D-Δ4,5,-glucuronopyranosyl (β1→3) *N*-acetyl-D-galactosamine 4-sulfate (D-gluco-4-enepyranosyluronic acid (β1→3)-2-acetamido-2-deoxy-D-galactose 4-sulfate; hereafter referred to as Δ4,5-disaccharide 4-sulfate), and the corresponding 6-sulfate (hereafter referred to as Δ4,5-disaccharide 6-sulfate, Figure 12). Both of these disaccharides could serve as substrates for the sulfatase activity of cell extracts and if this finding is considered in relation to the earlier studies of Dodgson and Lloyd[112] it implies that saturated depolymerization products (testicular hyaluronidase) or unsaturated products (chondroitinase) are equally acceptable as sulfatase substrates.

Strictly speaking, the mechanism of action of the chondroitinase of *P. vulgaris* is that of a lyase. However, the name chondroitinase will be retained for much of the present account because of its predominance in the relevant literature. It will later become apparent that four different chondroitinases are known, only two of which are listed in the 1978 edition of the recommendations of the Enzyme Nomenclature Committee. The recommended names of the listed enzymes will be cited at appropriate points in the text.

Some time elapsed before further studies by Suzuki and his colleagues[119] showed that cells grown on a peptone/yeast extract/salts medium (apparently without added chondroitin sulfate) actually produced two distinct sulfatases. One of these (chondro-4-sulfatase, EC 3.1.6.9) was active toward Δ4,5-disaccharide 4-sulfate and the other (chondro-6-sulfatase, EC 3.1.6.10) toward the corresponding 6-sulfate. *N*-Acetyl-D-galactosamine 4- or 6-sulfates and 4- or 6-sulfated saturated hexa-, penta-, tetra-, and trisaccharides derived from either chondroitin 4- or 6-sulfates by digestion with testicular hyaluronidase and (for the odd numbered oligosaccharides) β-glucuronidase, were not substrates for either enzyme. It will be apparent later that the sulfated saturated tetrasaccharide from chondroitin 4-sulfate can serve as a substrate for the 4-sulfatase in spite of the conclusions to the contrary by Suzuki's group.

In a further study,[120] the 4- and 6-sulfatases were each purified some 20- to 25-fold from extracts of cells grown in the presence of chondroitin 6-sulfate. Nucleic acid was removed from extracts by treatment with streptomycin sulfate and this was followed by fractionation with ammonium sulfate and separation of the two enzymes on DEAE-cellulose. Several additional points of importance emerged from this study. First, the

I.

II.

III.

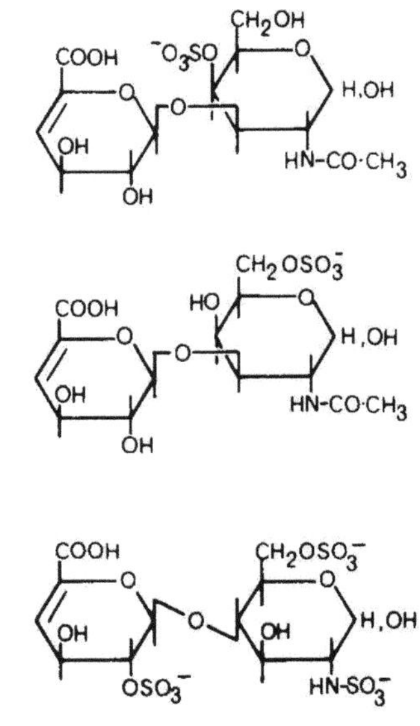

FIGURE 12. Unsaturated disaccharide products of the action of the chondroitinase-ABC of *Proteus vulgaris* on chondroitin 4- and 6-sulfates and dermatan sulfate, and the action of the heparin lyase of *Flavobacterium heparinum* on heparin. I, product from either chondroitin 4-sulfate or dermatan sulfate; II, product from chondroitin 6-sulfate; III, product from heparin.

activity of both sulfatases was greatly enhanced by the presence of chondroitin 6-sulfate in the medium. Second, the saturated sulfated disaccharides, *N*-acetylchondrosin 4- and 6-sulfates, were hydrolysed by the 4- and 6-sulfatases, respectively, but the 4- and 6-sulfate esters of *N*-acetyl-D-galactosamine were not. Surprisingly, however, the sulfate group at C-6 of *N*-acetyl-D-galactosamine 4,6-disulfate was removed by the 6-sulfatase. Maximum activity of both sulfatases was generally obtained at a pH around 7.5 in the presence of a Tris/HCl/acetate buffer mixture. For reasons that are not clear, the activity of the 4-sulfatase was enhanced (threefold) by the presence of acetate in the buffer mixture. Finally, it is worth noting that the 4-sulfatase seems to be a somewhat unstable enzyme.

Further information concerning substrate specificity came from an investigation[121] of the ability of the two sulfatases to remove sulfate from some oversulfated Δ4,5-unsaturated disaccharides isolated from digests of squid cartilage chondroitin sulfate, and dermatan sulfate isolated from bovine lung and pig skin. The first of these disaccharides, Δ4,5-disaccharide 4,6-disulfate, carried two sulfate groups on the *N*-acetyl-D-galactosamine residue and served as a substrate for both sulfatases, each sulfate group being removed by its own specific enzyme. The oversulfated disaccharide from dermatan sulfate carried the extra sulfate group at either C-2 or C-3 of the unsaturated uronic acid residue. The 4-sulfatase removed the 4-sulfate group from the hexosamine residue as expected, but the sulfate group on the unsaturated uronic residue was resistant to the enzyme as well as to the 6-sulfatase.

It would be useful at this juncture to devote some attention to the chondroitinase activity of *P. vulgaris*, as an understanding of its specificity is essential to the biodegradation story. Early studies[116-118] had established that hyaluronic acid, chondroitin

4- and 6-sulfates, and dermatan sulfate could be depolymerized by cell extracts, and an enzyme having activity for all four polymers was later purified to homogeneity. The production of the chondroitinase was enhanced some 150-fold by including chondroitin 4- or 6-sulfate in the growth medium. The enzyme is generally refered to as chondroitinase-ABC, the initials being taken from the older terms employed to describe chondroitin 4-sulfate (chondroitin sulfate A), dermatan sulfate (chondroitin sulfate B), and chondroitin 6-sulfate (chondroitin sulfate C). However, the name now recommended by the Enzyme Nomenclature Committee is chondroitin ABC lyase (EC 4.2.2.4). The relative activity of chondroitinase-ABC toward various substrates decreases in the order chondroitin 4-sulfate > chondroitin 4-sulfate-proteoglycan > chondroitin 6-sulfate > dermatan sulfate > hyaluronic acid.[122] The enzyme has since proved to be a most valuable tool in research on connective tissue proteoglycans.

Returning to the 4- and 6-sulfatases, it will be apparent that only two such enzymes are required for the biodegradation of chondroitin 4- and 6-sulfates and for dermatan sulfate. The role of the sulfatases in the biodegradation of the first two polymers is obvious, but it will be appreciated that the introduction, during depolymerization by the chondroitinase, of a double bond at the 4,5-positions of the L-iduronic acid residues of the dermatan sulfate (see Figure 6 for repeating unit), will eventually yield the same sulfated Δ4,5-unsaturated disaccharide as that produced from chondroitin 4-sulfate. The final desulfation stage in the biodegradation of dermatan sulfate is, therefore, the responsibility of the 4-sulfatase.

One final and important point about the sulfatases must now be considered. It will be recalled that Dodgson and Lloyd[112] had shown that a chondroitinase-free sulfatase preparation was able to liberate sulfate from the products of the digestion by testicular hyaluronidase of tracheal cartilage chondroitin sulfate (mainly 4-sulfate). The principal product of this digestion is a sulfated tetrasaccharide (Figure 13,I) and only trace amounts of the corresponding disaccharide are formed. This now needs to be considered in relation to the claim by Suzuki's group that sulfated tetrasaccharides prepared in this way from chondroitin 4- or 6-sulfate were not substrates for either the 4- or 6-sulfatases. This apparently contradictory situation his since been resolved by some elegant studies by Anno and her colleagues.[123] Briefly, a series of sulfated tetrasaccharides (see Figure 13 for structures) were isolated from hyaluronidase digests of chondroitin 4- and 6-sulfates and the chondroitin sulfates of squid and king crab cartilage. Tetrasaccharides I and II were the ones normally expected from the depolymerization of chondroitin 4- and 6-sulfates, respectively; tetrasaccharide III was also separated from a chondroitin 6-sulfate digest and contained one 6-sulfate group only, the reducing hexosamine moiety being sulfated at C-4; tetrasaccharide IV was from a squid cartilage digest and contained three sulfate groups, the nonreducing hexosamine having sulfate at C-6 as well as at C-4 as expected; tetrasaccharide V, obtained from king crab cartilage, had an additional ester sulfate group at C-2 or C-3 of the terminal glucuronic acid residue in addition to the expected 4-sulfate groups on the hexosamines. These various tetrasaccharides were then tested for their ability to serve as substrates for *P. vulgaris* 4- and 6-sulfatases and the results of these experiments are also shown in Figure 13. In summary, the 6-sulfatase is unable to remove the 6-sulfate group from any tetrasaccharide containing such a group. The 4-sulfatase readily removes the C-4 sulfate group located on the reducing terminal hexosamine, other sulfate groups being removed only very slowly. Finally, neither sulfatase was able to remove the sulfate group present at C-2 or C-3 of a terminal glucuronic acid residue. These findings confirm that tetrasaccharide I, almost certainly the major component present in the substrate mixture used by Dodgson and Lloyd,[112] can indeed serve as a sulfatase (4-sulfatase) substrate, and also the finding[119] of Suzuki's group that the corresponding tetrasaccharide prepared from chondroitin 6-sulfate (tetrasaccharide II) cannot. The failure of the group to detect desulfation of tetrasaccharide I remains a mystery.

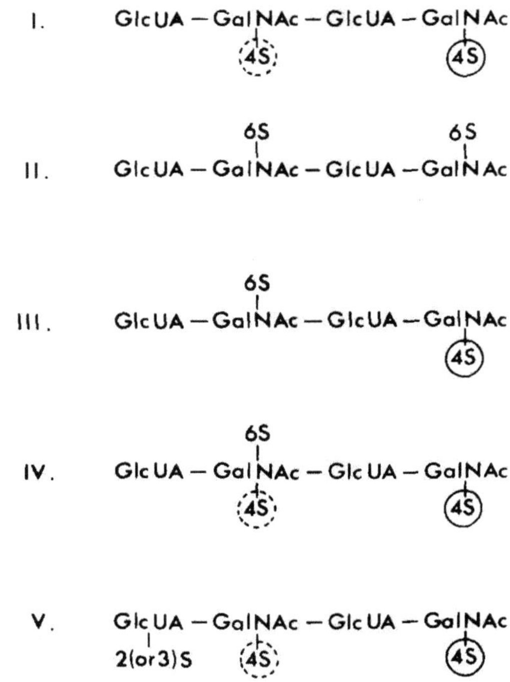

FIGURE 13. Action of the chondro-4- and 6-sulfatases of *Proteus vulgaris* on the tetrasaccharide products of the action of testicular hyaluronidase on chondroitin sulfate preparations from various sources. GlcUA = glucuronic acid; GalNAc = *N*-acetyl-D-galactosamine; 6S or 4S = position of sulfate group. Sulfate groups that are readily removed are circled with a solid line; those that are less readily removed are circled with a broken line, and those that are not removed are not circled. See text for the origins of the tetrasaccharides.

Little else remains to be said about the two *P. vulgaris* sulfatases, other than that the names conferred upon them by the Enzyme Nomenclature Committee (chondro 4- and 6-sulfatases) are not entirely satisfactory in the light of their established specificities. However, they serve well enough for the moment.

C. Chondrosulfatase Activity in *Flavobacterium heparinum*

During the course of a search for microbial systems that might be used in structural studies on heparin, Payza and Korn[124] isolated from soil by enrichment culture a bacterium that was able to utilize heparin as the sole source of carbon. The bacterium, to which the name *F. heparinum* was assigned, has since proved to be an extremely important tool in structural investigations on heparin and heparan sulfate and has also been of considerable importance to the general field of connective tissue biochemistry. The role of sulfatases produced by the organism in the biodegradation of heparin will be discussed at length in Section VII, but mention needs to be made now of enzyme systems produced by the bacterium that can participate in the degradation of chondroitin sulfate isomers. Meanwhile, it should be noted that a very recent study[125] on the morphology, physiology, biochemistry, and antimicrobial properties of this bacterium has indicated that it should be more properly assigned to the genus *Cytophaga* and the name *C. heparina*[125] (comb. nov.) has been suggested as appropriate. In the present work the name *F. heparinum* will be retained as it occurs in all the other literature cited in this chapter. The organism is maintained as number 13125 in the American Type Culture Collection.

Following the isolation of *F. heparinum* it was noted that after growth in a trypticase-phytone-glucose medium in the absence of heparin, cell extracts of the unadapted bacterium were still capable of degrading hyaluronic acid and chondroitin 4- and 6-sulfates even though they were unable to degrade heparin. Both sulfated and nonsulfated unsaturated oligosaccharides were produced, indicating the presence of constitutive sulfatase activity in the extracts. Dermatan sulfate, on the other hand, could only be degraded extremely slowly by the unadapted organism unless the cells were grown on a medium containing this polymer. Again sulfated and nonsulfated unsaturated disaccharides were obtained, indicating the presence of sulfatase activity. In a later study from the same laboratories it was further shown that the yields of the constitutive enzyme system acting on chondroitin 4- and 6-sulfates could be greatly enhanced if chondroitin 4-sulfate was present in the growth medium. The two unsaturated sulfated disaccharides obtained from digests of the 4- and 6-sulfates, respectively, appeared to be identical with those produced by extracts of *P. vulgaris*. Reincubation of the sulfated unsaturated disaccharide prepared from *F. heparinum* digests of chondroitin 4-sulfate with more cell extract led to the appearance of nonsulfated disaccharide, reaffirming the presence of sulfatase activity. One difference noted by the investigators between the sulfatase activity of cell extracts of nonadapted *F. heparinum* and the corresponding activity of cell extracts of *P. vulgaris* was that the latter system only was inhibited by phosphate.[112]

A more detailed study[126] of the chondroitinase and sulfatase activities of *F. heparinum* revealed that the position was somewhat more complex than was first apparent. According to the investigators, chondroitinase activity toward dermatan sulfate could also be enhanced if the organism was grown in the presence of chondroitin 6-sulfate as an alternative to dermatan sulfate and, under these circumstances, the adapted cells contained a chondroitinase active toward dermatan sulfate and chondroitin 4- and 6-sulfates, as well as a chondroitinase active only toward the last two of those compounds and toward hyaluronate. The activity of the latter enzyme was not enhanced when the bacterium was grown in the presence of dermatan sulfate. The first of these two systems appeared to be similar to *P. vulgaris* chondroitinase-ABC, while the second one is now generally referred to as chondroitinase-AC (chondroitin AC lyase, EC 4.2.2.5). No further work was done on the first system but the AC system was purified to apparent homogeneity and shown to be inactive toward dermatan sulfate, but to degrade chondroitin 4- and 6-sulfates to the same Δ4,5-unsaturated disaccharides as those produced by *P. vulgaris* chondroitinase-ABC. Chondroitinase-AC is now widely used in association with the *Proteus* ABC enzyme in studies on the biochemistry of connective tissues.

The sulfatase activity present in crude extracts of *F. heparinum* grown in the presence of chondroitin 6-sulfate presents some intriguing features. As mentioned earlier, phosphate does not inhibit the activity (cf. *P. vulgaris* sulfatases) but, more significantly, no enzyme appeared to be present that could desulfate directly the Δ4,5-disaccharide 6-sulfate. Sulfatase activity toward the corresponding 4-sulfate was present, but was not studied further.

In the event, subsequent investigations by Michelacci and Dietrich,[127-130] using electrophoresis as a means of resolving the different enzymes, showed that the pattern of chondroitinase activity produced by *F. heparinum* was somewhat more complex than had previously been envisaged. To summarize these studies, the chondroitinase-AC emerges as a constitutive enzyme, while growth of the bacterium in the presence of dermatan sulfate or chondroitin 4- or 6-sulfates leads to the production of a chondroitinase-B active only toward dermatan sulfate, and a chondroitinase-C active toward chondroitin 6-sulfate and hyaluronic acid. Neither enzyme is listed by the Enzyme Nomenclature Committee. Interestingly, chondroitinase-B is also very readily induced by either the Δ4,5-disaccharide 4- or 6-sulfates, with induction occurring much more rapidly than is the case when the polymers are used as inducers. The implication may be that a sulfated Δ4,5-unsaturated disaccharide is the true inducer of the enzyme, with the actual location of the sulfate group being unimportant.

Unfortunately, the studies of Michelacci and Dietrich on the chondroitinase have been of no help in furthering our knowledge of the potential of *F. heparinum* to produce sulfatases active toward dermatan sulfate and chondroitin 4- and 6-sulfates. Recently a new method was devised[131] for separating chondroitinases B and C by affinity chromatography on either a heparin-bound AH-Sepharose® 4B column coated with dermatan sulfate (for chondroitinase-B) or a dermatan sulfate-bound column coated with the same polymer (for chondroitinase-C). Sulfatase activity was followed during the course of these separations but, unfortunately, the selected assay substrate was Δ4,5-disaccharide 4-sulfate, so that an opportunity was missed to check for the presence of a 6-sulfatase analogous to that present in *P. vulgaris.*

How then to reconcile the early reports[116,126], that cell extracts of the unadapted bacterium were able to produce nonsulfated Δ4,5-unsaturated disaccharides from both chondroitin 4- and 6-sulfates with the failure of Suzuki's group[120] to detect a sulfatase active toward Δ4,5-disaccharide 6-sulfate? Two obvious explanations spring to mind; first, the 6-sulfatase is unstable and has therefore escaped detection; second, the 6-sulfate group is removed only after the hydrolysis of the Δ4,5-disaccharide 6-sulfate by the Δ4,5-disaccharide glucuronidase known to be present in adapted cells of *F. heparinum.*[116,120] This would leave the sulfate to be removed from the liberated *N*-acetyl-D-galactosamine 6-sulfate by a monosaccharide "glycosulfatase"-type of enzyme (see Section II). No recent attempts have been made to resolve these contradictory findings and, indeed, the fact that they are contradictory seems to have escaped notice. There, for the moment, the matter rests.

D. Chondrosulfatase Activity in *Bacteroides thetaiotaomicron*

For a variety of reasons, including some of commercial origin, increasing attention is being paid to the intestinal flora as a vehicle for the modification of chemical structures of compounds in transit. As part of a major study of the ability of colonic microorganisms to utilize polysaccharide material, Salyers and her colleagues[132] have spent some time investigating the ability to degrade chondroitin sulfate of one such organism, *B. thetaiotaomicron.* This Gram-negative, obligately anaerobic bacterium is a common isolate from human feces and strain VP15482A (NCTC 10852) was selected for detailed study.

Bacteria were grown in a defined medium (oxygen-free CO_2 atmosphere) containing a chondroitin 4-sulfate preparation (later shown to contain about 20% of the 6-sulfate) as the carbon source. Growth was paralleled by the disappearance of the polymer. Cell-free extracts were able to degrade chondroitin 4- and 6-sulfates as well as dermatan sulfate and contained chondroitinase and sulfatase systems. The former system, separated from the latter, gave the expected Δ4,5-disaccharide 4-sulfate as the product of the degradation of chondroitin 4-sulfate and dermatan sulfate, while the corresponding 6-sulfate was obtained as the degradation product of chondroitin 6-sulfate. Whether this implies the presence of a single chondroitinase-ABC or a multiplicity of enzymes (cf. *P. vulgaris)* remains to be established.

The sulfatase system of *B. thetaiotaomicron* was able to liberate sulfate from Δ4,5-disaccharide 4- and 6-sulfates and although it seems likely that two distinct sulfatases are present in the bacterium (cf. *P. vulgaris),* no evidence is available on this point.

A number of other very interesting findings concerned the cellular localization of the systems and it was shown that the chondroitinase activity was periplasmic, but was not an integral protein of either cell membrane. In contrast, the sulfatase activity, together with a β-glucuronidase active toward desulfated Δ4,5-disaccharide, resided in the cytoplasm proper. These findings directly established for *B. thetaiotaomicron* what was deduced from indirect evidence in the cases of *P. vulgaris* and *F. heparinum* namely, that extracellular enzymes played no part in the degradation of the chains of the chondroitin sulfate isomers. The implications of this are considered in a later section of this chapter.

In a further study from Salyer's laboratory[133] particular attention was paid to the induction of chondroitinase activity in *B. thetaiotaomicron*. The organism was grown in continuous culture in a defined medium that was limiting for glucose, the sole carbon source. Cultures were equilibrated for 7 to 10 generations before being transferred to stoppered incubation tubes (anaerobic conditions) containing chrondroitin 4-sulfate. The weak but detectable chondroitinase activity of glucose grown cells was enhanced more than tenfold within 90 min after the commencement of the induction process. It was possible to show that the synthesis of both RNA and protein was an essential feature of the oxygen-sensitive induction process but, in contrast to the situation in *F. heparinum* where chondroitinase-B activity could be induced by Δ4,5-disaccharide 4-sulfate, neither this compound nor the corresponding 6-sulfate could induce chondroitinase activity in *B. thetaiotaomicron*. However, these sulfated, unsaturated disaccharides, when tested as inducers, were not taken up into the cells so that the problem here might be one of transport. Attempts to induce the chondroitinase by sulfated saturated oligosaccharides prepared from testicular hyaluronidase digests of chondroitin 4-sulfate showed that oligomers containing eight or more residues were inducers with those containing ten or more residues being as effective as undegraded polymer.

The investigators gave no indication in their report as to whether sulfatase activity was subjected to similar induction events and it is to be hoped that this might be a subject of further work.

E. Other Possible Sources of Chondrosulfatase Enzymes

Searches for sources of enzymes that can participate in the degradation of chondroitin and dermatan sulfates continue to be made. Two principal reasons for such searches seem to be operating; first, from the point of view of clinical medicine and dentistry and, second, because of the desire to find other useful sources of chondroitinase activity that could be exploited for use as structural and analytical tools as has been the case for chondroitinases AC and ABC.

In relation to the medical aspect, it is possible that the presence of microorganisms that are able to degrade these polymers and their parent proteoglycan complexes may explain to some degree some clinical lesions that continue to defy precise interpretation. For example, the *Bacteroides* have recently become a focus of attention as a common cause of anaerobic infections in which the characteristic lesions produced are deep seated abscesses that may continue to invade the tissues.[134,135] It is not clear what factors do facilitate this degenerative process, but the possibility that dispersal of connective tissue matrix by the concerted action of a chondroitinase/chondrosulfatase system could be one of them.

The investigators involved in the various studies have generally employed the plate-method devised by Smith and Willett[108] for screening for "chondroitin sulfatase-producing" microorganisms. In the test, bacteria are grown on a solid medium consisting of Brain Heart Infusion broth to which is added Noble agar, chondroitin sulfate (bovine nasal cartilage), and bovine albumin (fraction V). After growth, the plates are flooded with 2 *N*-acetic acid for 10 min, when any nondegraded chondroitin sulfate precipitates as a complex with albumin. A clear zone is left around those colonies that produce soluble enzymes that can depolymerize the chondroitin sulfate.

In fact, the plate-method does not indicate chondrosulfatase production but only the ability of the microorganism to depolymerize the polymer so that it no longer precipitates as an albumin complex. Only a chondroitinase-type enzyme is required for this to happen. Although in the cases of *P. vulgaris, F. heparinum,* and *B. thetaiotaomicron,* sulfatase and chondroitinase activity coexist, there is no guarantee that this necessarily applies to other microorganisms. Unfortunately, a number of reports dealing with screening seem not to have appreciated these points.

Furthermore, because the method relies on the production of a zone of degraded chondroitin sulfate around each active colony, it is reasonable to assume that significant extracellular release of enzyme must occur. It will be recalled that the chondroitinase/chondrosulfatase activities of *P. vulgaris, F. heparinum,* and *B. thetaiotaomicron* are apparently localized within the cell and there is no certainty in such cases that sufficient enzyme would be released extracellularly to allow a positive result in the plate-method. Care is therefore needed in interpreting screening reports as indicating the presence or absence of "chondroitin sulfatase-producing" microorganisms. This is not to say that the plate-method is not a most valuable addition to the methodology available for the study of glycosaminoglycan-degrading organisms. It is important, however, to remember that positive or negative results do not necessarily mean the presence or absence of sulfatase activity.

In the original study of Smith and Willett[108] strains of *Corynebacterium acnes* and *Propionibacterium acnes* gave positive tests for "chondroitin sulfatase-producing" activity, and in a later study[136] 22 strains of *P. acnes* and two strains of *Propionibacterium granulosum* gave strong positive tests for such activity. Rudek and Haque[137] also used the plate-method to screen for extracellular "chondroitin sulfatase" and other degradative enzyme activities in various species or subspecies of *Bacteroides.* With one exception, all of the subspecies of *B. fragilis* were active. The same investigators later purified an extracellular chondroitinase-type enzyme from the cell-free growth medium of *B. distasonis.*[138] The enzyme preparation was also active toward hyaluronate, but no attempts were apparently made to check for sulfatase activity. Other workers[109,110,134-141] have also employed the plate technique for similar types of study but without apparently realizing that they were observing chondroitinase activity, but not necessarily "chondroitin sulfatase" activity.

Turning now to the search for enzyme systems that may be exploited for use as analytical and structural tools, the work of Kitamikado and Lee[142] stands out for special mention. In conducting the search, the investigators recognized that the plate-method did not reveal sulfatase activity. Moreover, they also realized the other limitations of the method and modified it so that colonies that grew on solid media containing chondroitin 4-sulfate were transferred to liquid media containing polymer. After a growth period of 3 days, an acidified albumin solution was added and failure to obtain a chondroitin 4-sulfate-albumin complex indicated chondroitinase-producing organisms. This modification clearly extended the chance of detecting enzyme activity that was not readily secreted into the growth medium. A total of 253 chondroitinase-producing bacteria were obtained from aquatic and land sources, the majority being assigned to seven genera, namely, *Aeromonas, Arthrobacter, Beneckea, Flavobacterium, Micrococcus, Proteus* and *Vibrio.* A number of active isolates were examined for intracellular and extracellular chondroitinase production and the results clearly indicated (Table 2) that some organisms produced large quantities of extracellular enzyme (e.g., *Aeromonas* species), whereas others produced very little (e.g., *Proteus* and *Flavobacterium* species). Finally, a chondroitinase-AC enzyme was purified (977-fold) in good yield from the culture fluid of *Aeromonas* sp. 83.

Table 2
SURVEY OF BACTERIAL CHONDROITINASE ACTIVITY ACCORDING TO KITAMIKADO AND LEE[142]

Strain (no.)	Bacterium	Intracellular	Extracellular	Bacterium isolated or obtained from
118	*Proteus vulgaris*	2.83[a]	0.02	Gills of flat fish *Limanda herzensteini*
127	*P. vulgaris*	2.38[a]	0.01	Intestine of mackerel *Scomber japonicus*
129	*P. vulgaris*	2.06[a]	0.02	Stomach of mackerel *S. japonicus*
126	*P. vulgaris*	1.57[a]	0.01	Intestine of mackerel *S. japonicus*
105	*Flavobacterium balustinum*	1.15[b]	0.13	Marine mud
145	*P. vulgaris*	0.88[a]	0.02	Feces of raccoon dog *Nyctereutes procyonoides*
137	*F. diffusum*	0.39[b]	0.01	Marine mud
139	*Beneckea labra*	0.27[b]	0.01	Marine mud
325	*B. chitinovora*	0.24[b]	0.17	Green alga *Codium fragile*
83	*Aeromonas* sp.	0.75[b]	16.68[b]	Intestine of carp *Cyprinus carpio*
326	*Aeromonas* sp.	0.11	14.70[b]	Feces of coatimundi *Nasua narica*
123	*Aeromonas* sp.	0.13	6.61[b]	Intestine of carp *C. carpio*
302	*Aeromonas* sp	0.06	5.91[b]	Intestine of carp *C. carpio*
144	*Aeromonas* sp.	0.05	5.40[b]	Feces of raccoon dog *N. procyonoides*
84	*Aeromonas* sp.	0.02	4.29[b]	Intestine of carp *C. carpio*
303	*Aeromonas* sp.	0.06	3.85[b]	Intestine of carp *C. carpio*
125	*Micrococcus varians*	0.10	2.94[b]	Intestine of eel *Anguilla japonica*
14	*Vibrio* sp.	0.05	2.58[b]	Intestine of mackerel *S. japonicus*
6	*Vibrio* sp.	0.01	2.52[b]	Intestine of mackerel *S. japonicus*
	F. heparinum	0.52[a]	0.06	ATCC 13125
	P. vulgaris	1.22[a]	0.02	NCTC 4636

Note: Enzyme activity was assayed using chondroitin 6-sulfate and was expressed as units present in a 20 m*ℓ* culture. One unit was defined as liberating μmol of Δ4,5-glucuronysyl-*N*-acetylgalactosamine-6-sulfate per minute.

[a] Chonroitinase — ABC activity.
[b] Chondroitinase — AC activity.

Unfortunately, the authors of this otherwise excellent work, failed to examine the selected isolates for sulfatase activity and an opportunity was lost to throw further light on the relationship between the two systems as far as coexistence is concerned.

Before closing this particular section, the attention of the reader is drawn to a series of papers[143-147] on an extracellular chondroitinase-AC produced by a soil isolate identified as *Arthrobacter aurescens*, and to studies on a chondroitinase-ABC produced by *P. mirabilis*.[148] The first of these systems has been purified[143] from the culture fluid following growth of *A. aurescens* in the presence of chondroitin 6-sulfate. The properties of the enzyme have been described[144] and it has been shown that the mode of action is somewhat different from that of the corresponding enzyme from *F. heparinum*.[145] Furthermore, three forms of *A. aurescens* chondroitinase-AC can be separated by isoelectric focusing. None of the reports provide any indication whether sulfatase enzymes coexist with the chondroitinase.

In the case of the *P. mirabilis* system, the ABC enzyme was induced by growth in the presence of chondroitin sulfate and was purified some 400-fold from cell extracts. The enzyme, which did not appear to be secreted into the medium, showed activities toward chondroitin 6- and 4-sulfates and dermatan sulfate which decreased in that order. Attempts to detect extracellular or intracellular sulfatase activity were unsuccessful.

F. Concluding Remarks

Much remains to be accomplished concerning the ability of microorganisms to produce sulfatases that can participate in the biodegradation of chondroitin 4- and 6-sulfates and dermatan sulfate. In spite of the failure to detect sulfatase activity associated with the chondroitinase-ABC of *P. mirabilis*, the fact that an appreciable number of different microorganisms from a variety of sources are able to produce some form(s) of chondroitinase leads one to suspect that the distribution of chondrosulfatases may be more widespread than is currently apparent. There is an urgent need to examine this possibility further.

In the case of an organism such as *P. vulgaris* NCTC 4636 it seems clear that the overall degradation of chondroitin and dermatan sulfates occurs intracellularly rather than extracellularly, and that it involves an initial depolymerization to either Δ4,5-disaccharide 4- or 6-sulfate. Desulfation by the appropriate 4- or 6-sulfatase follows, yielding the same Δ4,5-unsaturated disaccharide, D-Δ4,5-glucuronopyranosyl (β1→3)-*N*-acetyl-D-galactosamine. A β-glucuronidase enzyme[116,120,132] specific for Δ4,5-unsaturated disaccharide and which is inactive toward the β1→3 D-glucuronosidic bond of saturated disaccharides then hydrolyses the unsaturated disaccharide to yield *N*-acetyl-D-galactosamine and the Δ4,5-unsaturated uronic acid. Both compounds can then presumably serve as carbon sources for growth although no one appears to have checked this.

It will be recalled that chondroitinase and chondrosulfatase systems in *P. vulgaris*, *F. heparinum*, and *B. thetaiotaomicron* are located within the cell. Sulfated glycosaminoglycan chains can therefore presumably penetrate the outer membrane of the cells of these bacteria. If the situation in *B. thetaiotaomicron* is typical of all three organisms then the glycosaminoglycan chains would encounter the depolymerizing enzymes in the periplasmic region. The Δ4,5-disaccharide 4- and 6-sulfates resulting from depolymerization would then penetrate the inner membrane where they would encounter the sulfatase and β-glucuronidase enzymes.

An entirely different set of circumstances must operate in bacteria such as *A. aurescens* that produce their chondroitinases extracellularly (see also Table 2). Presumably, the glycosaminoglycan chains cannot penetrate the outer membrane and must therefore be depolymerized by extracellular enzymes. The subsequent fate of the liberated Δ4,5-disaccharide 4- and 6-sulfates remains a mystery for the present. Logically, one might expect that they would be more readily accessible to the cell proper where sulfate and β-glucuronidase action would complete the depolymerization process. This lack of precise information emphasizes the need for further studies on the sulfatase and β-glucuronidase enzymes, and hinders attempts to rationalize the ecological significance of the difference between those bacteria that depolymerize substrate extracellularly (when products may also become available to other organisms) and those that depolymerize substrate intracellularly.

Of course, under natural circumstances sulfated glycosaminoglycan chains may well be presented to microorganisms as components of proteoglycans or proteoglycan aggregates; in which case their utilization for growth purposes might be presumed to involve the preliminary activity of one or more proteases. A number of the bacteria that produce extracellular chondroitinase activity also secrete proteolytic enzymes into the culture medium.[136,137] It would now be of considerable interest to see how an organism such as *P. vulgaris* or *B. thetaiotaomicron* would deal with a proteoglycan complex. Are extracellular proteases part of the enzyme complement of such organisms or are intact proteoglycans able to penetrate the outer cell membrane?

As a final comment, it should be stressed that the commercial availability of chondroitinases ABC and AC, and chondro-4- and 6-sulfatases has made a considerable impact on studies of connective tissues and their macromolecular components. Space does not permit an account of the uses to which the enzymes have been put, but it will perhaps suffice to indicate their potential for the determination of small quantities of individual components of polymer mixtures,[149] the determination of structural divergency from the norm in glycosaminoglycan preparation from different sources,[123] the elucidation of the glycosaminoglycan composition of tissues,[150] the assay of urinary glycosaminoglycans in patients with metabolic disorders involving connective tissues,[151] and for investigations into growth and differentiation.[152]

VI. KERATAN SULFATASES

Two distinct sulfatases appear to be involved in the biodegradation of keratan sulfate by microorganisms. In common with microbial sulfatases participating in the biodegradation of carrageenan and the isomeric chondroitin sulfates (see Sections IV and V), the keratan sulfatases show no activity toward the intact polymer, but rely on an initial depolymerization of the polymer to provide their substrates.

Keratan sulfate chains are widely distributed, in association with the chondroitin sulfates, as components of proteoglycan aggregates in cartilaginous and skeletal structures. They also occur uniquely as the sole glycosaminoglycan component of a proteoglycan present in the corneal tissue of the eye. It will be recalled from Section I.A.5.c. of this chapter that keratan sulfate chains consist for the most part of a disaccharide repeating unit of D-galactose (β1→4)-*N*-acetyl-D-glucosamine with sulfate always present at C-6 of the hexosamine, but sometimes present at C-6 of the hexose (see Figure 6). These repeating units are 1→3 linked to each other. The reader is reminded that this represents an oversimplified view of the structure of the polymer and the review by Roden[42] should be consulted for additional details.

Comparatively few searches for microbial enzymes that can participate in the biodegradation of keratan sulfate chains seem to have been made, but it is worth noting at the outset that bacteria that have been studied for their ability to degrade the isomeric chondroitin sulfates do not have the ability to degrade keratan sulfate. This is in spite of the fact that keratan sulfate and chondroitin sulfate chains are frequently found attached to the same protein core in a proteoglycan. Such searches that have been made have been prompted by the belief that availability of isolated enzymes of known specificity that can participate in the degradation of keratan sulfate might help to clarify polymer structure in those regions of the chains which depart from the norm. Four microbial sources of such enzymes are currently known.

A. Sulfatase and Other Enzyme Activities in *Coccobacillus*

Meyer and his colleagues[153] were the first to find a bacterium that would degrade keratan sulfate. Cell extracts of the organism, a *Coccobacillus* species (ATCC 13949), degraded the polymer to a mixture of oligosaccharides, D-galactose, *N*-acetyl-D-glucosamine, and inorganic sulfate, thereby indicating the participation of some sort of sulfatase as well as some form of depolymerase.

Subsequently, the organism was grown at 30° C with minimal aeration in a basal salts medium containing purified gastric mucin and a depolymerizing enzyme was partially purified and freed from the associated sulfatase and exoglycosidase activities.[154] When keratan sulfate preparations from either bovine cornea or human costal cartilage were incubated with the enzyme, a series of sulfated oligosaccharides was obtained. One of the major oligosaccharide components was a tetrasaccharide containing two equivalents each of *N*-acetyl-D-glucosamine and D-galactose, and three equivalents of sulfate. Later work[155] saw the enzyme purification procedure improved (113-fold purification but still heterogeneous) and the demonstration that D-galactose was present at the reducing end of all the oligosaccharides formed as a result of its action. Chemically desulfated keratan sulfate preparations were degraded at a faster rate than the parent polymer. This was taken to indicate that the enzyme, considered to be an endo-β-D-galactosidase, was partly inhibited by the presence of sulfate groups. In spite of the possible implications of this finding in terms of the role of sulfatase activity in the biodegradation of keratan sulfate, no further attempt was made to investigate that activity and work on the bacterium seems to have terminated at that juncture.

B. Sulfatase and Other Enzyme Activities in *Escherichia freundii*

Concurrently with the *Coccobacillus* studies of Meyer's group, Japanese investigators were searching for other bacteria that could degrade keratan sulfate.[156] In this search keratan sulfate prepared from whale cartilage was always included in the growth medium. Twenty-five laboratory species of aerobic and anaerobic bacteria of the genera *Pseudomonas, Escherichia, Proteus, Lactobacillus,* or *Clostridium* were examined, as well as organisms present in ten soils, four pond waters, two river waters, two sea waters, two air dusts, and human and dog feces. Ability to degrade keratan sulfates prepared from either cornea or whale cartilage was restricted to the fecal organisms, and then only if the polymer had been present in the growth medium. A bacterium, later identified[157] as a strain of *Escherichia freundii,* was isolated and grown aerobically in the presence of keratan sulfate. Keratan sulfate-degrading activity was present in cell extracts as well as in the culture medium. *N*-Acetyl-D-glucosamine, D-galactose, and inorganic sulfate were found as degradation products. Clearly, a sulfatase of some type was involved; moreover, the extracts contained an *N*-acetylhexosaminidase and a β-galactosidase. Subsequently,[158] a depolymerizing enzyme was purified (1,750-fold) from the culture fluid but no attempts were apparently made to isolate and study the sulfatase activity of the bacterium.

The strain of *E. freundii* subsequently passed into the hands of other investigators who noted[159] that the depolymerizing enzyme was actually an endo-β-galactosidase that degraded keratan sulfate to oligosaccharides in which D-galactose was present as terminal residue at the reducing end of each one. The enzyme would also hydrolyse a sulfated glycoprotein prepared from pig colonic mucin and small oligosaccharides such as lacto-*N*-tetraose (D-galactose[β1→3]-*N*-acetyl-D-glucosamine[β1→3]-D-galactose[β1→4]-D-glucose).

The endo-β-galactosidase was later[160] purified 4,400-fold from the culture fluid with a recovery of 45%. The pure enzyme hydrolysed keratan sulfate from a variety of sources in a random fashion to produce oligosaccharides of various molecular sizes. Structural analysis of a disaccharide and a tetrasaccharide isolated from the enzyme digests showed the former to be *N*-acetyl-D-glucosamine 6-sulfate (β1→3)D-galactose (Figure 9,I) and the latter to be *N*-acetyl-D-glucosamine 6-sulfate (β1→3)D-galactose 6-sulfate (β1→4)-*N*-acetyl-D-glucosamine 6-sulfate (β1→3)-D-galactose (Figure 9,II). These figures imply that the endo-β-galactosidase specifically attacks those endo-β-galactosidic bonds in which nonsulfated D-galactose residues participate. In accord with this view it was noted that keratan sulfates with a higher content of sulfated galactose (e.g., shark cartilage keratan sulfate) were hydrolysed to a lesser extent and yielded oligosaccharides of a relatively large size compared to bovine keratan sulfate. Herein also lies the probable explanation for the findings by Hirano and Meyer[155] (see previous section) that the *Coccobacillus* endo-β-galactosidase was particularly active toward chemically desulfated keratan sulfate.

During the course of the studies on *E. freundii* an opportunity was taken to investigate the sulfatase activity of cell extracts and the culture fluid. Activity was measured toward keratan sulfate, D-galactose 6-sulfate and *N*-acetyl-D-glucosamine 6-sulfate. No activity toward any of the substrates could be detected in the culture fluid, but activity toward all three was observed with extracts. In contrast, about 80% of the total endo-β-galactosidase activity was present in the culture fluid. No comments were made by the investigators on the possible role of the sulfatase activity in the overall biodegradation of keratan sulfate.

C. Sulfatase and Other Enzyme Activities in *Pseudomonas* and *Actinobacillus* Species

The reason for linking these two organisms together will soon become apparent. Both came into prominence as the result of a search by Suzuki's group for new microorganisms that might produce enzymes that could aid studies on the chemical variability of keratan sulfates. A number of soil bacteria of unspecified type were examined,[88] but the only two isolates showing any promise were *Pseudomonas* IFO-13309 and *Actinobacillus* IFO-13310. Both could grow in defined liquid media containing shark keratan sulfate as the sole source of carbon but, interestingly, the growth rate achieved when both organisms were present in the same culture flask greatly exceeded that of either when grown separately. Some form of symbiotic relationship between the two organisms was obviously occurring, possibly reflecting the sharing of different enzymes between the two.

Chromatograms of the products of prolonged digestion of corneal keratan sulfate by cell extracts of *Pseudomonas* IFO-13309, when grown independently, revealed the presence of a number of sulfated oligosaccharides of varying molecular size, whereas analogous *Actinobacillus* IFO-13310 digests contained mainly D-galactose and *N*-acetyl-D-glucosamine. The investigators considered that these differences could be explained, at least in part, by assuming that the *Pseudomonas* extract was particularly rich in endoglycosidase activity, but poor in exoglycosidase activity, whereas the opposite situation prevailed in the *Actinobacillus* extracts. It was clear from the chromatograms of digests that sulfatase activity must be present in the latter extracts, but not clear that any such activity was present in *Pseudomonas* extracts. In fact both organisms possess such activity and the release of sulfate by *Pseudomonas* extracts during digestion of keratan sulfate could readily be measured. Presumably, these apparently conflicting findings could be reconciled if it was assumed, in the case of *Pseudomonas*, that only a part of the sulfate of any oligosaccharide was removed, and if exoglycosidase activity was in some way related to sulfatase activity in the case of *Actinobacillus* extracts.

Some of these suppositions were established as fact by the findings that *Actinobacillus* extracts had high β-*N*-acetylglucosaminidase activity, relatively high exo-β-galactosidase activity, but low endoglycosidase activity. This situation was reversed in *Pseudomonas* extracts which were high in endoglycosidase activity and very low in the exoglycosidase activities. Sulfatase activity toward a sulfated oligosaccharide preparation seemed to be very similar in both extracts. As a passing comment, and in the light of the work on *E. freundii*, one is tempted to wonder whether some of the endoglycosidase activity of both bacteria was being secreted into the culture fluid during growth, but no account of this possibility appears to have been taken. It is rather interesting that the growth rate of *Pseudomonas* IFO-13309 was increased tenfold if the *Actinobacillus* was also present, whereas the growth rate of the latter organism only doubled if the pseudomonad was present.

In order to make further progress, the investigators undertook the purification of the *Actinobacillus* sulfatase activity and a 200-fold purification of the enzyme from cell extracts was achieved by a procedure involving removal of nucleic acid with streptomycin sulfate, fractionation with ammonium sulfate, and chromatography on CM-cellulose and Sephadex® G-200. During the course of purification there was evidence of the presence of a second and distinct sulfatase and further comment on this enzyme will be made later.

Meanwhile, the purified sulfatase was tested for its ability to liberate sulfate from a series of naturally occurring or synthetic sulfate esters. Sulfate was released from *N*-acetyl-D-glucosamine 6-sulfate, the disaccharide *N*-acetyl-D-glucosamine 6-sulfate (1→3)-D-galactose, and from the tetrasaccharide *N*-acetyl-D-glucosamine 6-sulfate (1→3)-D-galactose 6-sulfate (1→4)-*N*-acetyl-D-glucosamine 6-sulfate (1→3)-D-galactose. However, only one third of the ester sulfate present was released from the tetrasaccharide and this was shown to represent the selective hydrolysis of the sulfate attached at C-6 to the terminal, nonreducing *N*-acetyl-D-glucosamine residue. The investigators concluded that a single enzyme, an *N*-acetyl-D-glucosamine 6-sulfatase, was responsible for all three activities. *p*-Nitrophenyl sulfate, D-galactose 3- and 6-sulfates, *N*-acetyl-D-galactosamine 4- and 6-sulfates, Δ4,5-disaccharide 4- and 6-sulfates (see Section V.B) keratan sulfate, chondroitin 4- and 6-sulfates, dermatan sulfate, heparan sulfate, and heparin were not substrates for the enzyme.

Turning briefly to the finding that a second sulfatase appeared to be present in crude cell extracts and in some fractions obtained during the purification of the *N*-acetyl-D-glucosamine 6-sulfatase, preliminary experiments indicated that it was active toward D-galactose 6-sulfate, but not toward *N*-acetyl-D-glucosamine 6-sulfate.

Collectively, therefore *Actinobacillus* IFO-13310 possesses the following enzymes relevant to the degradation of keratan sulfate:

1. An endo-β-galactosidase (only present in low amounts in cell extracts, presence in culture fluid never assessed).
2. A sulfatase active toward *N*-acetyl-D-glucosamine 6-sulfate.
3. An exo-β-*N*-acetylglucosaminidase.
4. A sulfatase active toward D-galactose 6-sulfate.
5. An exo-β-galactosidase.

Further comment on this will be deferred.

During further studies Nakazawa and Suzuki[161] succeeded in purifying from cell extracts the endo-β-galactosidase of *Pseudomonas* IFO-13309. The purified enzyme was then used to degrade keratan sulfate. A number of oligosaccharides differing in size and sulfate content were obtained and the two major components were characterized. From these and other results it was possible to conclude that the bond that was susceptible to the enzyme was associated with the structure β-D-galactose(1→4)-*N*-acetyl-D-glucosamine 6-sulfate. This was deduced from the finding that the two major oligosaccharides in the enzyme digests contained D-galactose at the nonreducing end. Evidence was also obtained that the enzyme had little or no activity toward regions of the keratan sulfate chains that carried D-galactose residues bearing sulfate at C-6 (about 40% of the D-galactose residues in corneal keratan sulfate). These results agree with those obtained for the endo-β-galactosidase of *E. freundii*[160] (see Section VI.B).

A further important contribution to the whole story of sulfatase activity in relation to keratan sulfate degradation came from Nakazawa and Kagabe[89] who purified from extracts of *Actinobacillus* IFO-13310 the sulfatase with activity toward D-galactose 6-sulfate that had been noted earlier.[88] An 81-fold purification (yield, 6.8%) was obtained by a procedure involving removal of nucleic acid, fractionation with ammonium sulfate, and chromatography on CM-cellulose, DEAE-Sephadex® A-25, Sephadex® G-200 and Sephacryl® S-200. The final preparation was free from the associated *N*-acetyl-D-glucosamine 6-sulfatase and was active toward *N*-acetyl-D-galactosamine 6-sulfate as well as D-galactose 6-sulfate (ratio of activities, 2.4:1). The enzyme also removed the sulfate group from the terminal D-galactose 6-sulfate residue at the nonreducing end of a trisaccharide (see Figure 9,III) prepared from keratan sulfate by sequential degradation with endo-β-galactosidase, *N*-acetyl-D-glucosamine 6-sulfatase, and exo-β-*N*-acetylglucosaminidase. The sulfatase was inactive toward *p*-nitrophenyl sulfate, *N*-acetyl-D-galactosamine 4-sulfate, *N*-acetyl-D-glucosamine 6-sulfate, *N*-acetyl-D-galactosamine 3,4- and 4,6-disulfates, D-glucose 6-sulfate, Δ4,5-disaccharide 6-sulfate (see Section V.B.), keratan sulfate, and chondroitin 6-sulfate.

Interestingly, during the course of the purification of this enzyme, the presence of a third sulfatase active toward D-glucose 6-sulfate was noted. This enzyme also exhibited feeble activity toward *N*-acetyl-D-glucosamine 6-sulfate. This enzyme would appear to have no role to play in the biodegradation of keratan sulfate and has not been studied further.

D. Concluding Remarks

The results obtained with the *Coccobacillus, E. freundii, Pseudomonas* IFO-13309, and *Actinobacillus* IFO-13310 may now be considered as a whole. Collectively they provide good, although not unequivocal, evidence of the sequence of events in the biodegradation of keratan sulfates.

First, an endo-β-galactosidase is a common feature of all four bacteria and in two cases *(E. freundii* and *Pseudomonas* IFO-13309) it is known that the enzyme depolymerizes the substrates by hydrolysing the β1→4 galactosidic bond between D-galactose and *N*-acetyl-D-glucosamine 6-sulfate residues, providing that the D-galactose carries no sulfate group. Somewhat less direct evidence indicates a similar system is operating in *Coccobacillus* ATCC 13949. By inference, it is not unduly unrealistic to assume that the endo-β-galactosidase found in relatively low amounts in *Actinobacillus* exhibits similar behavior. In three cases *(Coccobacillus, E. freundii,* and *Actinobacillus)* prolonged incubation of cell extracts with keratan sulfate leads to the production of *N*-acetyl-D-glucosamine, D-galactose, and inorganic sulfate as the major end-products, indicating that exo-β-*N*-acetylglucosaminidase, exo-β-galactosidase and sulfatases must be involved. The presence of such enzymes has been demonstrated in both *E. freundii* and *Actinobacillus,* and in both cases sulfatase activities toward *N*-acetyl-D-glucosamine 6-sulfate and D-galactose 6-sulfate are present. In the latter organism, two different sulfatases are known to be involved.

Taken into account with other results detailed in Section VI.A and B, these findings now permit a sequence of events to be proposed for the biodegradation of keratan sulfates:

1. Endo-β-galactosidase depolymerizes the substrate to yield oligosaccharide fragments, each containing an *N*-acetyl-D-glucosamine 6-sulfate residue at the nonreducing end of the molecule.
2. The intervention of *N*-acetyl-D-glucosamine 6-sulfatase removes the sulfate group from each of the terminal residues exposed as a result of stage 1.
3. Exo-β-*N*-acetylhexosaminidase removes the desulfated hexosamine residue.
4. If the D-galactose residue exposed as a result of stage 3 possesses a sulfate group at C-6, that group is then removed by D-galactose 6-sulfatase.
5. Exo-β-D-galactosidase removes the desulfated D-galactose residue, thereby exposing a further *N*-acetyl-D-glucosamine 6-sulfate for the attention of the stage 2 sulfatase; and so on.

The fact that oligosaccharides of varying but still relatively small size are produced when corneal keratan sulfate is degraded by the endo-β-galactosidase,[161] indicates that unsulfated D-galactose residues must be distributed frequently, but unevenly along the polysaccharide chain. With increasing sulfation of these residues the whole process of biodegradation of keratan sulfate is greatly slowed and much larger oligosaccharide fragments are obtained. How effectively these larger fragments can serve as substrates for the *N*-acetyl-D-glucosamine 6-sulfatase and hence (subsequently) for the other exoglycosidases and D-galactose 6-sulfatase is completely unknown, but would form an important and interesting topic for further research.

Finally, the reader is reminded that in an earlier section of this chapter (Section II) the problem was raised of the use of the term "glycosulfatase" for enzymes that can use sulfated di-, tetra-, or oligosaccharides as substrates in addition to simple sulfated monosaccharides. The *Actinobacillus* sulfatases were cited as a case in point. Little needs be added to the comments made in that section other than to reiterate that logical and realistic names for such enzymes must await the availability of a great deal of additional information concerning their relative specificities. At present, sulfatases participating in keratan sulfate biodegradation do not figure in the Enzyme Nomenclature Committee lists.

VII. SULFATASES INVOLVED IN THE DEGRADATION OF HEPARIN AND HEPARAN SULFATES

Previous reference has been made (Section I.A.5.c, this chapter) to the complexity and heterogeneity of heparin and (particularly) heparan sulfates. In dealing with enzyme systems participating in the biodegradation of these two polymer types it will be necessary to take an oversimplified view of structures, but the reader is referred to the useful reviews by Rodén[42] Lindahl,[162] and Lindahl and Höök[163] for additional details.

For reasons that will shortly become obvious it is convenient to discuss the biodegradation of the two polymers under the same major heading so that features common to both can eventually be drawn together. However, parts of the account will be presented under separate subheadings.

A. Heparin-Degrading Enzymes

The trisulfated (and major) repeating unit of heparin (generally 70 to 80% of the total units) has been presented earlier (Figure 7) and will suffice for understanding the role of the sulfatase enzymes in the biodegradation process. However, an idealized and larger fragment of the structure, as shown in Figure 14 may help to remind the reader of structural variations that are possible and may assist him in following the account.

1. Historical Development

The story of heparin biodegradation by bacteria is really dominated by a single organism, *F. heparinum (Cytophaga heparinum,* see Section V.C). This bacterium was isolated from soil by enrichment culture on a liquid basal salts medium containing 0.1% of heparin.[124] The obligate aerobe grew optimally at 24° C and not at all at 0° or 37° C. Because of the difficulties and cost of obtaining quantities of heparin appropriate to the large scale growth of the bacterium, such growth was first achieved on a trypticase-phytone-glucose-salts medium containing only a small amount (0.002%) of heparin. The bacterium grew well, but was not able to degrade heparin to any significant extent. Consequently, harvested cells were washed with 25 m*M*-phosphate buffer, pH 8.0, and adapted by suspending them in that buffer (one tenth of the volume of the original culture fluid) containing 0.05% casein hydrolysate and 0.1% heparin, and aerating vigorously at 24° C for 2 to 4 hr. Phosphate was essential for the adaptation, but all salts seemed to inhibit the adaptation process if present at concentrations exceeding 30 m*M*.

In subsequent experiments[164-166] extracts of acetone-dried adapted cells were used to degrade heparin, the degradation being followed by determination of liberated reducing materials, decrease in metachromasia, periodate titration, and amino sugar release. It was concluded that several enzymes were involved, including some form of glycosidase (heparinase) able to break the glucosaminidic bond, a sulfamidase (sulfamatase) able to remove the *N*-sulfate (sulfamide, N−S bond) groups at position 2 of the hexosamine moieties (deduced because of the need for an unblocked C-2 amino group in

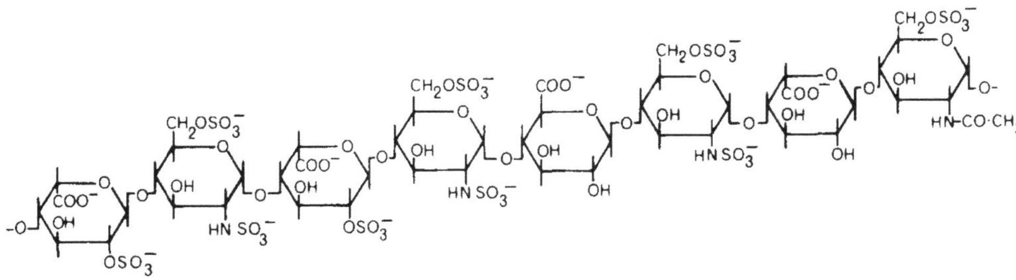

FIGURE 14. Idealized octasaccharide illustrating the structural variations that are possible in the heparin molecule.

the Elson-Morgan determination of amino sugar release), a glycosidase able to break the uronidic bond (higher levels of liberated reducing materials than expected from amino sugar release alone), and a sulfatase (deduced from the ratio of the increase in the periodate titration to the reducing groups formed). It was later shown[126] that chemically desulfated heparin was not degraded by cell extracts, suggesting that the presence of sulfate groups was necessary for the attack of the initial depolymerizing enzyme.

It will become apparent later that most of these early findings were subsequently shown to be substantially correct. Progress in elucidating the full details and the sequence of events was, however, slow. This probably reflected difficulties experienced in detecting and isolating the intermediate compounds produced during the degradative process. In addition, it must be appreciated that the chemical structure of heparin was quite uncertain at that time and, indeed, the principal objectives of the bacterial work was to learn more about that structure. A pointer to the way ahead came when Korn and Meyer's group agreed to collaborate in studies on the chondroitinase activity of *F. heparinum*[116,126] and established the role of that lyase enzyme in producing sulfated Δ4,5-unsaturated disaccharides from isomeric chondroitin sulfates (see Section V.C). It then became logical to think of heparin degradation in similar terms and eventually an enzyme fraction was separated from cell extracts and shown to possess an eliminase-type enzyme producing unsaturated oligosaccharides and a very active glycuronidase-type enzyme that apparently hydrolysed unsaturated disaccharides before they could be detected in significant amounts.[167] The products of the action of these enzymes on heparin were tentatively identified as an α-keto acid formed by ketonization of the Δ4,5-unsaturated uronic acid liberated by the glycuronidase, D-glucosamine *N*-sulfate (2-deoxy-2-sulfamido-D-glucose) and an *O*- and *N*-disulfated D-glucosamine.

At this juncture it seemed that progress would now be rapid but, in the event, some confusion developed following the entry of C. P. Dietrich into the field of research. In an initial study[168] of the ability of cell extracts of the same strain of *F. heparinum* to degrade heparin, Dietrich isolated a disulfated hexosamine from the enzyme digests and tentatively concluded it to be D-glucosamine 2,6-disulfate (2-deoxy-2-sulfamido-D-glucopyranosyl 6-sulfate). In addition, he isolated several sulfated oligosaccharides as well as two sulfated disaccharides, one containing three sulfate groups and the other containing two such groups. These he chose to identify as saturated disaccharides ("3-sulfatoglucuronylglucosamine 2,6-disulfate" and "glucuronylglucosamine 2,6-disulfate", respectively) rather than the unsaturated disaccharides proposed by the other research group. This implied that an endoglycosidase and not an eliminase (lyase) was responsible for the depolymerization step.

Dietrich retained his view of the matter in subsequent studies in which he separated and partially purified a sulfamidase and a "sulfoesterase" (disulfogucosamine 6-sulfatase, EC 3.1.6.11),[169] a "glucosaminidase" and a "glycuronidase",[170] and a "uronic acid isomerase" that could convert glucuronic acid to fructuronic acid.[171] On the basis of these findings, it was proposed that heparin was degraded by an endoglucosaminidase to yield either trisulfated or disulfated disaccharides, depending on the particular structure of the heparin chain at the point of origin of the disaccharides. Any disulfated disaccharide was hydrolysed by the glycuronidase to yield D-glucosamine 2,6-disulfate and uronic acid. The 6-sulfate group of the former was then removed by the 6-sulfatase to give D-glucosamine *N*-sulfate which could then serve as a substrate for the sulfamidase. The alternative possibility that these last two enzymes should act in the reverse order was also suggested. It was less clear how the trisulfated disaccharide fitted into the picture, but it seems to have been assumed that some sort of sulfatase might convert it to the disulfated disaccharide. No specific role was assigned to the "uronic acid isomerase."

Contemporary with the final paper in this particular Dietrich series, Hovingh and Linker[172] returned to the main point at issue, namely the mechanism of the depolymerization step, and reinforced their original conclusion that the enzyme was an eliminase (generally called heparinase; recommended name, heparin lyase, EC 4.2.2.7) producing sulfated Δ4,5-unsaturated disaccharides. Shortly afterwards Dietrich and his colleagues[173] accepted this view of the matter and provided good evidence that the principal disaccharide produced was the trisulfated Δ4,5-unsaturated disaccharide, 2-sulfo-L-Δ4,5-iduronopyranosyl (α1→4)-D-glucosamine 2,6-disulfate (2-sulfo-L-ido-4-enepyranosyluronic (α1→4)-2-sulfamido-2-deoxy-D-glucose 6-sulfate). This compound will be referred to hereafter as trisulfated Δ4,5-hep. disaccharide (Figure 12). Further supportive evidence for this conclusion came from Linker's group[174,175] which was successful in separating from cell extracts an α-glycuronidase enzyme that hydrolysed the disulfated Δ4,5-unsaturated disaccharide, L-Δ4,5-iduronopyranosyl (α1→4)-D-glucosamine 2,6-disulfate (hereafter referred to as disulfated Δ4,5-hep. disaccharide). The liberated uronic acid was relatively unstable and equilibrated with the open chain form, thereby releasing a hydroxyl group that was suitably positioned for a keto-enol transformation in which the keto form predominated.

Almost all the necessary information was now available to allow a scheme to be presented that would account for the biodegradation of the greater part of the heparin molecule, and the final link in the chain came when the Dietrich group[176] established the existence in cell extracts of a "disaccharide sulfoesterase" (hereafter referred to as Δ4,5-uronate sulfatase) that removed the sulfate group from the C-2 position of the unsaturated uronic acid residue. Using agarose-gel electrophoresis, these investigators went on to separate (and partially purify) the heparin lyase, the Δ4,5-uronate sulfatase and the α-glycuronidase from a fraction containing both the sulfamidase and the disulfoglucosamine 6-sulfatase. Collectively, the five enzymes were able to degrade the greater part of the heparin chain to its basic components, and the use of appropriate recombinations of some of the enzymes allowed the sequence of events to be established with reasonable certainty. Figure 15, which presents the Dietrich view of this sequence of events (including alternatives), probably needs no additional comment other than to say that the scheme applies to the (major) part of the heparin molecule containing the trisulfated disaccharide repeating unit (Figure 7) and that the Δ4,5-uronate sulfatase seems to be the rate-limiting enzyme in the sequence.

2. The Participating Sulfatases

a. Δ4,5-Uronate Sulfatase

This enzyme was purified about fivefold by agarose-gel electrophoresis and separated from other enzymes that could participate in heparin degradation.[176] Activity toward arbitrary concentrations of substrate was proportional to the time of incubation and to the concentration of enzyme. A pH of 7.0 was generally employed for enzyme experiments, but it is not clear whether this is the optimum pH. The enzyme was shown to be a simple hydrolase active only towards the trisulfated Δ4,5-hep. dis-

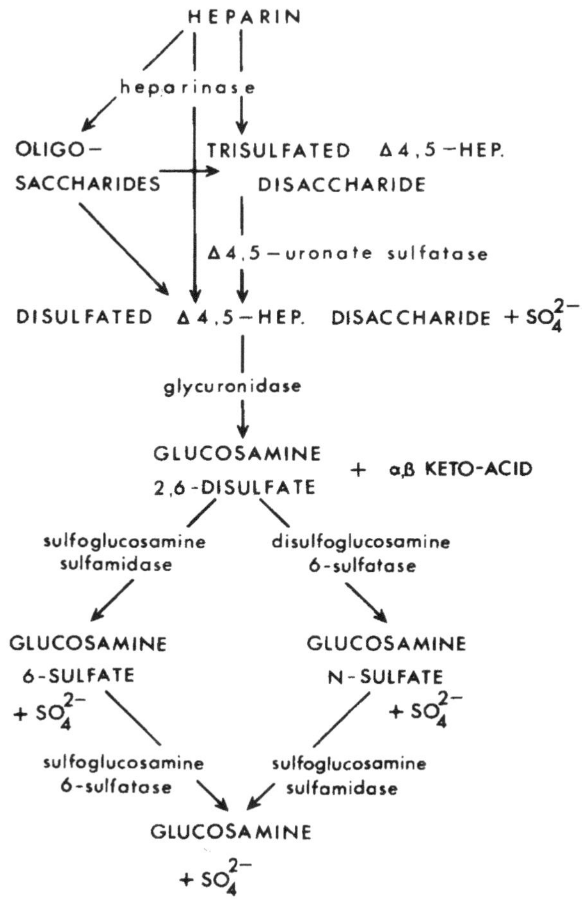

FIGURE 15. Scheme (modified from Dietrich et al.[176]) illustrating the probable sequence of events (including alternatives) in the biodegradation of the major part of the heparin molecule by the enzyme systems of *Flavobacterium heparinum*.

accharide. Heparin, disulfated Δ4,5-hep. disaccharide, D-glucosamine 2,6-disulfate, and sulfated hexa- and tetrasaccharides obtained by the action of heparin lyase on heparin, were not substrates. Mn^{2+}, Ca^{2+}, Mg^{2+}, Na^{+}, and K^{+} had no effect on the enzyme, but 10^{-3} *M* Ba^{2+}, Co^{2+}, or Fe^{2+} inhibited activity completely. This sulfatase is not listed by the Enzyme Nomenclature Committee.

b. Disulfoglucosamine 6-Sulfatase

The first experiments[169] on this enzyme were carried out with a pellet fraction separated from cell extracts of *F. heparinum* by centrifuging at 100,000 g for 1 hr. The bulk of the 6-sulfatase activity resided in the pellet whereas the heparin lyase remained in the supernatant. The sulfatase was optimally active toward D-glucosamine, 2,6-disulfate at 45° C and pH 7.0. There was some indirect evidence to suggest that D-glucosamine 6-sulfate could also serve as substrate for the enzyme. Phosphate inhibited and Mg^{2+} activated the enzyme, the K_m of which was 8×10^{-5} *M*. Subsequently, the sulfatase was purified 8-fold by agarose-gel electrophoresis, but was still contaminated with the sulfamidase.[176] It was devoid of activity toward heparin or toward trisulfated and disulfated Δ4,5-hep. disaccharides. No further studies of any consequence seem to have been made.

c. Sulfoglucosamine Sulfamidase

Initial studies on this enzyme[169] were made with the same 100,000 g pellet described in the last section. Unlike the disulfoglucosamine 6-sulfatase the sulfamidase was not inhibited by phosphate. The optimum pH was in the region of 7.0, the optimum temperature for activity was 25° C and the K_m value obtained under those conditions with D-glucosamine 2,6-disulfate as substrate was 1.3×10^{-4} *M*. The enzyme was also active towards D-glucosamine *N*-sulfate, but little or no other work by the investigators seems to have been done with that substrate. Agarose-gel electrophoresis was used[176] to obtain partially purified preparation which still contained the 6-sulfatase, but there is no record of any further work on this preparation. Some additional information on the enzyme came from a study[177] on a partially purified enzyme preparation that still contained traces of heparin lyase. Heparin, disulfated Δ4,5-hep. disaccharide, D-glucosamine *N*-sulfate, and D-glucosamine 2,6-sulfate, all of which carried ^{35}S-labeled *N*-sulfate groups, were tested as substrates. The last three were *N*-desulfated by the enzyme preparation. However, in view of the absence of information concerning the possible levels of α-glycuronidase in the preparation and the presence or absence of phosphate in the incubation mixtures (phosphate inhibiting any disulfoglucosamine 6-sulfatase activity) it can only be concluded with certainty that D-glucosamine *N*-sulfate was a true substrate. Activity toward that substrate was maximal at pH 7.2 the optimum substrate concentration was 16×10^{-3}m*M* and the K_m value was 5.3×10^{-3} *M*. This value is an order of magnitude greater than that obtained by Dietrich (1.3×10^{-4} *M*) when D-glucosamine 2,6-disulfate was used as substrate, and this tends to support the view that the action of the sulfamidase may precede that of the disulfoglucosamine 6-sulfatase during heparin biodegradation. Interestingly, L-serine *N*-sulfate, D-cycloserine *N*-sulfate, and phenylsulfamate did not inhibit sulfamidase activity toward D-glucosamine *N*-sulfate[177] and further comment on this point is made in Volume II, Chapter 1.

B. Heparan Sulfate-Degrading Enzymes

It was mentioned in Section I.A.5.c of this chapter that heparan sulfates really constitute a family of molecules that are closely related to heparin. Two of the major repeating units are shown in Figure 7, II, III, but nonsulfated as well as trisulfated residues are also known to occur. Most of the work on the biodegradation of this polymer has been less concerned with basic enzymological problems than with the need to obtain enzymes that could assist in unraveling the complexities of the families of heparan sulfates and, because of the close structural similarities to heparin, biodegradation studies on both polymers have tended to go hand in hand.

1. Historical Development

It was clear from an early date[126,178] that extracts of *F. heparinum* cells that had been adapted either to heparin or heparan sulfate (former name heparitin sulfate) were capable of degrading both polymers. Following this finding, heparan sulfate degradation formed the subject of a series of papers from Linker's group[167,172,179,180,181] which culminated in the discovery that extracts of adapted cells contained, in addition to heparin lyase, a heparan sulfate lyase (heparitinase in most of the literature, heparitin sulfate lyase, EC 4.2.2.8) that participated in the depolymerization of heparan sulfate, yielding products containing Δ4,5-unsaturated uronic acid residues. The heparin lyase seemed to require the presence of either *N*-acetyl or *N*-sulfate groups for its activity, but also the absence of *O*-sulfate groups in the vicinity of the glycosidic bond to be ruptured. Using partially purified preparations of the heparin lyase and heparan sulfate lyase enzymes it was further shown[182] that a total of five different disaccharides could be separated from digests of various heparan sulfate fractions that had been isolated[182] from lung and aorta. These disaccharides included

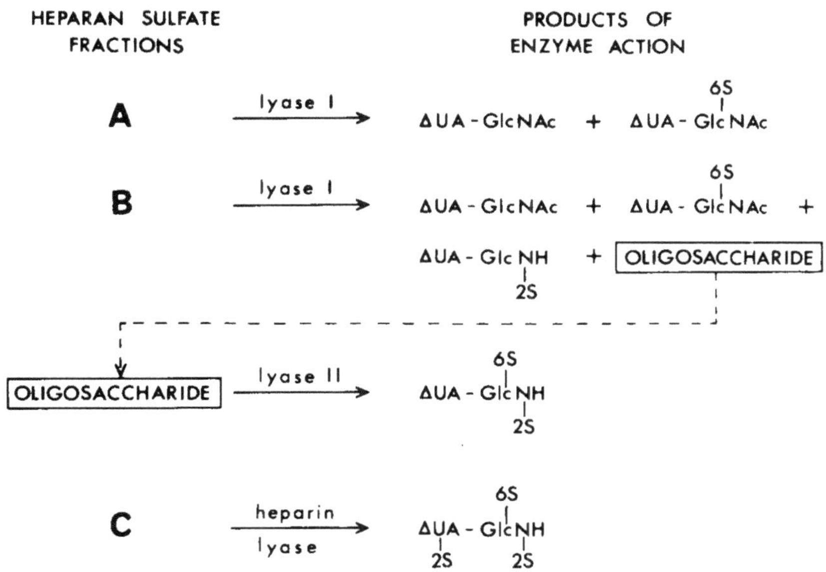

FIGURE 16. Schematic representation of the action of heparan sulfate lyases I and II (lyases I and II) and heparin lyase on three heparan sulfate fractions separated from a heterogeneous commercial sample of the polymer. Abbreviations used to describe the unsaturated disaccharide products are: ΔUA = Δ4,5-unsaturated uronic acid; GlcNAc = *N*-acetyl-D-glucosamine; GlcNH = D-glucosamine; 2S and 6S = positions of sulfate groups.

1. The same trisulfated Δ4,5-hep. disaccharide that constitutes the principal repeating unit of heparin.
2. A disulfated disaccharide apparently unique to heparan sulfates and containing a sulfated uronic acid residue linked to D-glucosamine *N*-sulfate.
3. A monosulfated disaccharide containing a D-glucosamine *N*-sulfate residue.
4. A monosulfated disaccharide containing *O*-sulfated *N*-acetyl-D-glucosamine.
5. A nonsulfated disaccharide containing *N*-acetyl-D-glucosamine.

The first two of these were produced when heparin lyase was used as the depolymerase, whereas the other three were products of heparan sulfate lyase action. These results certainly confirmed the structural complexity of heparan sulfates and supported the view that was gaining credence at the time, that heparin-like fragments were present in the polymer.

Collectively, the results of the various investigations established that heparin lyase could degrade some portions of heparan sulfate chains, whereas heparan sulfate lyase exhibited only relatively poor activity toward heparin.

At this point in the historical sequence of events, the Dietrich group[183,184] made the further interesting discovery that extracts of cells adapted to either heparin or heparan sulfates actually contained two different heparan sulfate lyase enzymes (heparan sulfate lyases I and II). These could be separated from each other and from heparin lyase by large scale electrophoresis in agarose gels. The three enzymes were then each tested for ability to degrade four distinct heparan sulfate fractions (A, B, C, and D) that had been separated by polyacrylamide-gel electrophoresis of commercial samples of the polymer. One of these (heparan sulfate D) was subsequently revealed as a contaminating fragment of heparin and will not be mentioned further. The results of the enzyme digestion experiments are presented in idealized form in Figure 16, from which it can be seen that the complete conversion of heparan sulfates A, B, and C to disaccharide components would require the presence of all three lyases.

One further point that might be made before leaving the heparan sulfate lyase enzymes is that lyase II is able to participate in those (limited) regions of the heparin molecule in which trisulfated repeating units are replaced by disulfated repeating units.[185] The Enzyme Nomenclature Committee lists do not appear to recognize the existence of two heparan sulfate lyases.

2. Participating Sulfatases

Little needs to be said about the participation of sulfatases in the biodegradation of heparan sulfates by *F. heparinum*. The various sulfated unsaturated disaccharide products of the combined action of heparin lyase and heparan sulfate lyases I and II bear sulfate groups at one or more positions that enable them to serve as substrates (either before or after glycuronidase action) for one or more of the same three sulfate liberating enzymes that participate in heparin biodegradation. These enzymes, Δ4,5-uronate sulfatase, disulfoglucosamine 6-sulfatase and glucosamine sulfamidase have already been described in detail earlier in this chapter (Section VII.A.2).

C. Induction of Heparin- and Heparan-Sulfate-Degrading Enzymes

It is now time to return to the valuable original studies that emanated from Korn's laboratory, and to take up a point that was made concerning the induction of the *Flavobacterium* enzyme systems involved in heparin biodegradation. Comparative studies on the degradation of heparin, hyaluronate, and chondroitin sulfate by preparations of unadapted and adapted cells led Korn[166] to conclude that the only enzyme concerned in the adaption was the sulfamidase and this enzyme was therefore presumed to be the one that acted first on heparin, removing *N*-sulfate groups which perhaps inhibited the heparin lyase. Later work on the system clearly negated this interpretation of the degradative sequence. However, studies by Lloyd et al.[186] showed that growth of *F. heparinum* on either heparin or D-glucosamine *N*-sulfate, both possessing ^{35}S-labeled sulfamido groups, led to the production of sulfamidase activity toward both substrates. Similar results were obtained if the cells were adapted with either the polymer or the monomer, so that it was possible that a small molecule, rather than a polymer, might be the actual inducer.

This problem was further investigated by Dietrich[187] who concluded that the inducer might be a sulfated disaccharide formed from heparin by the action of constitutive enzymes present in minute quantities in nonadapted cells. Subsequently, a large number of relevant compounds were tested for their ability to act as inducers, with all results being related to that obtained with a standard heparin sample.[188] The compounds tested included the following polymers: heparan sulfates A, B, C, and D (see Section VII.B.1), *N*-desulfated heparin, *N*-,*O*-desulfated-*N*-reacetylated heparin, and the corresponding methyl ester, chondroitin 4- and 6-sulfates, dermatan sulfate, molluscan heparin, and pectin. A number of sulfated and unsulfated Δ4,5-unsaturated disaccharides prepared by bacterial degradation of chondroitin 4- and 6-sulfates and heparin and fungal degradation of pectin were also tested, as were the monosaccharides, D-glucosamine 2, 6-disulfate, D-glucosamine *N*-sulfate, D-glucuronic acid, and L-iduronic acid.

All the compounds tested induced heparin-degrading activity, but heparan sulfates were especially active (more than twice as effective as heparin) as were trisulfated Δ4,5-hep. disaccharide and disulfated Δ4,5-hep. disaccharide (1.3 and 1.1 times as effective, respectively). The fact that heparan sulfate A, which contains very little ester sulfate, was equally as good as heparin as an inducer indicated that the presence of sulfate is not essential for the induction of the heparin-degrading enzymes and that *N*-acetyl groups can substitute for *N*-sulfate groups. It was concluded that the minimal structural requirements for a satisfactory inducer are the presence of a uronic acid and an *N*-sulfated or *N*-acetylated hexosamine linked together by an $\alpha 1 \rightarrow 4$ glycosidic linkage.

Interesting as these experiments were, they were based on the production of D-glucosamine 2,6-disulfate by cells in the presence of 20 m*M* phosphate buffer (to inhibit disulfoglucosamine 6-sulfatase) and therefore had to involve the sequential action of heparin lyase (possibly heparan sulfate lyase II also), the Δ4,5-uronate sulfatase, and the α-glycuronidase. The position is further complicated by the suggestion[176] that the Δ4,5-uronate sulfatase is rate limiting in the overall degrading process. This pinpoints the difficulties inherent in induction experiments that do not consider all parts of a multienzyme system and there is clearly a great need for induction studies on the individual steps of the overall process. Meanwhile, it is probably reasonable to assume that heparan sulfate lyases I and II are coinduced with the heparin lyase and by the same inducers. However, this remains to be proven.

D. Concluding Remarks

At the beginning of the account of heparin degradation it was pointed out that a single bacterium dominated the story. However, at least one other bacterium is known to possess a similar capability. This organism, a *Bacteroides* species which was isolated by enrichment culture from the normal intestinal flora of man, was a strict anaerobe which grew poorly on normal media.[189] However, a medium was devised that produced satisfactory growth and it was shown that the bacterium grown in the absence of heparin was able to degrade chondroitin sulfate and hyaluronate, but not heparin. Introduction of heparin into the growth medium led to the appearance of heparin-degrading activity. Cell extracts of the organism degraded heparin as evidence by the disappearance of metachromasia and the appearance of reducing materials. Maximum degrading activity occurred at around pH 7.0 and at a heparin concentration of about 10 mg/mℓ. Degradation was accompanied by release of hexosamine end-groups as determined by the Elson-Morgan method and this indicates the presence of a sulfamidase as liberated hexosamines with a blocked C-2 amino group will not react in the method. No further studies on the enzymes of this bacterium seem to have been made.

We are therefore left with *F. heparinum* a truly versatile bacterium from the point of view of glycosaminoglycan biodegradation. Cell extracts of the unadapted bacterium are able to degrade hyaluronate and chondroitin 4- and 6-sulfates (see Section V.C) and, after appropriate adaptation, can also degrade dermatan sulfate, heparin, and heparan sulfates. This overall capacity involves production of a constitutive chondroitinase-AC, inducible chondroitinases B and C, heparin lyase, and heparan sulfate lyases I and II. Additionally, chondroitin 4-sulfatase, Δ4,5-uronate sulfatase, disulfoglucosamine 6-sulfate, glucosamine sulfamidase, and two distinct glycuronidases (one mainly specific for β-D-[1→3]-derived linkages and the other for β-D[1→4] and α-L-[1→4]-derived linkages)[190] also feature in appropriate cell extracts of the organism. Not surprisingly, the capacity to produce such a variety of enzymes has proved to be particularly important to studies on glycosaminoglycans. The commercial availability and importance as an analytical and structural tool of chondroitinase-AC, used in association with *P. vulgaris* chondroitinase-ABC, has been the subject of comment in an earlier section (V.F). Both these enzymes can be further coupled with heparin lyase and heparan sulfate lyase preparations and used for the identification of small amounts of heparin, heparan sulfates, and other tissue glycosaminoglycans.[191] Linker and Hovingh[192] have reviewed some of the ways in which the various enzymes can be used for identification and structural analysis of glycosaminoglycans and examples of such use are now encountered with increasing frequency.

Much of the account of heparin and heparan sulfate biodegradation has, of necessity, been concerned with depolymerizing enzymes. However, this book is concerned with sulfatase enzymes and, in the final analysis it seems that almost nothing is known about those involved in the breakdown of the two polymers. The manner of their induction, the precise intracellular localization; their general and kinetic properties; and their relationship to each other in terms of structure, binding, and catalytic sites and mechanisms of action, seem not to have concerned any investigator in spite of the fascinating challenge presented.

VIII. GENERAL COMMENTS ON CARBOHYDRATE SULFATASES

Further reference to these enzymes will be made in Volume II, Chapter 4, but a number of points are worthy of comment at this juncture. First, it must be apparent from consideration of the remarkable number and structural variability of the potential natural substrates mentioned in the introduction to this chapter that a vast field of research is awaiting exploitation. Reference was made to potential substrates with ester sulfate groups present on D-glucose, D- and L-galactose, 3,6-anhydro-D-galactose, L-rhamnose, L-fucose, D-mannose, D-xylose, D-arabinose, D-glucosamine, D-galactosamine, L-iduronic acid, and D-glycerol. Some of the sulfated residues mentioned are present in relatively simple structures, others in remarkably complex ones. Doubtless others yet remain to be discovered. At the present time we have no idea what sort of sulfatases will be involved in the biodegradation of 7 of the 13 sulfate-carrying monosaccharide residues cited, even ignoring the fact that the position of sulfate groups on the sugar molecule may also be very variable.

In considering the routes of biodegradation of these many and varied carbohydrate sulfate esters, it is certain that such compounds as the branched-chain seaweed polysaccharide sulfates or the sulfated glycoproteins present a most formidable challenge. Not only do new microorganisms have to be isolated, but the inevitable multiplicity of enzymes certain to be involved in biodegradation will make it difficult to pinpoint the precise role of sulfatases that are involved. However, even a simple disaccharide sulfate such as lactose 6-sulfate presents a challenge. Do bacteria exist that are able to remove the sulfate group directly from the disaccharide or must the disaccharide first serve as a glycosidase substrate before desulfation of the liberated glucose 6-sulfate can occur? Indeed, there are grounds for wondering (see Section III) if any liberated glucose 6-sulfate might first have to be converted to gluconate 6-sulfate or glycerate 3-sulfate before encountering a suitable sulfatase. What if the lactose 6-sulfate molecule is further complicated by the presence of an additional moiety as, for example, in neuramin lactose 6-sulfate? Endless possibilities present themselves through such musings and yet sufficient has already been done to point to the way in which the various problems might be tackled. For example, there are strong indications from the studies on κ-carrageenan and glycosaminoglycan biodegradation that depolymerization is likely to be a prerequisite for sulfatase action, and it seems probable that sulfatase substrates are likely to be relatively small molecules. Additionally, the studies on glycosaminoglycan biodegradation have indicated that the participating sulfatases may exhibit a relatively high specificity, so that one would certainly be foolish not to look for more than one participating sulfatase if the compound under investigation contained sulfate groups on different monosaccharide residues or at different positions on the same type of residue.

In searching for such biodegradation enzyme systems there has probably been a change of emphasis from the earlier approaches to enzyme separation that involved the laborious manipulation of chromatography columns of various sorts. The success achieved with the heparin- and heparan-sulfate-degrading enzymes by separating them on agarose-gel electrophoresis[183,184] perhaps points the way to rapid progress in the understanding of the number of enzymes involved and their sequence of action. However, interest in carbohydrate sulfatases has recently moved away from the realm of microorganisms to that of mammals, following the realization that the enzymes were involved (by their absence) in a number of serious inherited metabolic disorders. Space does not permit a discussion of the various enzymes that are involved, but the reader is referred to a recent review by Farooqui[193] for details. It now remains to be seen whether this transfer of interest will mean that the field of microbial carbohydrate sulfatases falls into neglect. This would be a great pity because so little has yet been done. Mention was made in Volume I, Chapter 1 that sulfatase enzymes have frequently emerged as particularly fascinating subjects of study for the enzymologists because of their curious kinetic properties, mechanisms of action, and occasional strange dual specificity toward apparently unrelated compounds. The reader will encounter a number of relevant examples in Volume II, Chapter 2. Additionally, microbial carbohydrate sulfatases should be of particular interest to soil microbiologists and biochemists because of the variety and quantity of their substrates that are daily returned to the environment. In the past, interest in soil sulfate esters and sulfatase activity has largely been confined to arylsulfatases and their substrates because of the ready availability of the latter and the simple methods that can be used to measure the former. However, the development of techniques and methodology have been such that no one need now be unduly concerned about embarking on what used to be considered as a difficult and perhaps not worthwhile field of scientific endeavor — the carbohydrate sulfatases.

REFERENCES

1. **Schweiger, R. G., Ed.,** *Carbohydrate Sulfates,* American Chemical Society Symposium Series, American Chemical Society, Washington, D.C., 1978.
2. **Mumma, R. O. and Verlangiera, J.,** Isolation of ascorbic acid 2-sulfate from selected rat organs, *Biochim. Biophys. Acta,* 273, 249, 1972.
3. **Suzuki, S. and Strominger, J. L.,** Enzymic sulfation of mucopolysaccharides in hen oviduct, *J. Biol. Chem.,* 235, 257, 1960.
4. **Stoffyn, P. J.,** The structure and chemistry of sulfatides, *J. Am. Oil Chem. Soc.,* 43, 69, 1966.
5. **Ishizuka, I., Suzuki, M., and Yamakawa, T.,** Isolation and characterization of a novel sulfoglycolipid, "seminolipid" from boar testis and spermatazoa, *J. Biochem. Tokyo,* 73, 77, 1973.
6. **Cooper-Driver, G. and Swain, T.,** Sulphate esters of caffeyl and p-coumaryglucose in ferns, *Phytochemistry,* 14, 2506, 1975.
7. **Harborne, J. B. and Mokhtari, N.,** Two sulfated anthroquinone derivatives in *Rumex pulcher, Phytochemistry,* 16, 1314, 1977.
8. **von Wyler, H., Rösler, H., Mercier, M., and Dreiding, A. S.,** Präbetanin, ein Schwefelsäuer-halbester des bétanins. Ein beitrag zur kenntnis der betacyane, *Helv. Chim. Acta,* 50, 545, 1967.
9. **Imperato, F.,** Betanin 3-sulphate from *Rivinia humalis, Phytochemistry,* 14, 2526, 1975.
10. **Liem, P. Q. and Laur, M.-H.,** Structures tenaues et compositions des esters sulfuriques, sulfoniques, phosphoniques des glycosyldiglycerides de trois fucacées, *Biochimie,* 58, 1367, 1976.
11. **Golden, M. and Hunter, A. S.,** Organic sulfate esters of potential biological interest, *J. Hetrocyclic Chem.,* 6, 739, 1969.
12. **Hancock, A. J. and Kates, M.,** Structure determination of the phosphatidylglycerosulfate (diether analog) from *Halobacterium cutirubrum, J. Lipid Res.,* 14, 422, 1973.
13. **Barra, H. S. and Caputto, R.,** Isolation and identification of a lactose sulphate ester from rat mammary gland, *Biochim. Biophys. Acta,* 101, 367, 1965.
14. **Ryan, L. C., Carubelli, R., Caputto, R., and Trucco, R. E.,** Studies on the structure of neuramin lactose sulfate, *Biochim. Biophys. Acta,* 101, 252, 1965.
15. **Martensson, E.,** On the sulfate containing lipids of human kidney, *Acta Chem. Scand.,* 17, 1174, 1963.
16. **Stoffyn, A., Stoffyn, P., and Martensson, E.,** Structure of kidney ceramide dihexoside sulfate, *Biochim. Biophys. Acta,* 152, 353, 1968.
17. **Goren, M. B., Brokl, O., Roller, P., Fales, H. M., and Das, B. C.,** Studies of *Mycobacterium tuberculosis:* the structure of the principal sulfatide (SL-I), *Biochemistry,* 15, 2728, 1976.

18. **Slomiany, B. L., Slomiany, A., and Horowitz, M. I.**, Sulfatides of hog gastric mucosa, *Biochim. Biophys. Acta,* 348, 388, 1974.
19. **Slomiany, B. L., Slomiany, A., and Glass, G. B. J.**, Glycolipids of the human gastric content: structure of the sulfated glyceroglucolipid, *Eur. J. Biochem.,* 78, 33, 1977.
20. **Kates, M. and Deroo, P. W.**, Structure determination of the glycolipid sulfate from the extreme halophile *Halobacterium cutirubrum, J. Lipid Res.,* 14, 438, 1973.
21. **Slomiany, A., Smith, F. B., and Slomiany, B. L.**, Isolation and characterization of a sulfated glyceroglucolipid from alveolar lavage of rabbit, *Eur. J. Biochem.,* 98, 47, 1979.
22. **Nigrelli, R. F., Chanley, J. D., Kohn, S. K., and Sobotka, H.**, The chemical nature of holothurin, a toxic principle from the sea-cucumber *(Echinodermata holothurioidea), Zoologica,* 40, 47, 1955.
23. **Yasumoto, T., Nakamura, K., and Hashimoto, Y.**, A new saponin, holothurin B, isolated from sea-cucumbers, *Holothuria vagabunda* and *Holothuria lubrica, Agric. Biol. Chem. (Tokyo),* 31, 7, 1967.
24. **Chanley, J. D. and Rossi, C.**, The neo-holothurinogenins-III: neo-holothurinogenins by enzymatic hydrolysis of desulfated holothurin A, *Tetrahedron,* 25, 1911, 1969.
25. **Medcalf, D. G.**, Sulfated fucose-containing polysaccharides from brown algae: structural features and biochemical implications, in *Carbohydrate Sulfates,* Schweiger, R. G., Ed., American Chemical Society Symposium Series, Washington, D.C., 1978, chap. 15.
26. **Percival, E.**, Sulfated polysaccharides of the Rhodophyceae — a review, in *Carbohydrate Sulfates,* Schweiger, R. G., Ed., American Chemical Society Symposium Series, Washington, D.C., 1978, chap. 14.
27. **Heaney-Kieras, J., Rodén, L., and Chapman, D. J.**, The covalent linkage of protein to carbohydrate in the extracellular protein-polysaccharide from the red alga *Porphyridium cruentum, Biochem. J.,* 165, 1, 1977.
28. **Percival, E. and Foyle, R. A. J.**, The extracellular polysaccharides of *Porphyridium cruentum* and *Porphyridium aerugineum, Carbohydr. Res.,* 72, 165, 1979.
29. **Percival, E.**, Sulfated polysaccharides metabolized by the marine chlorophyceae — a review,in *Carbohydrate Sulfates,* Schweiger, R. G., Ed., American Chemical Society Symposium Series, Washington, D.C., 1978, chap. 13.
30. **Mackie, W. and Preston, R. D.**, Cell wall and polysaccharides, in *Algal Physiology and Biochemistry,* Stewart, W. D. O., Ed., Blackwell Scientific, Oxford, 1974, chap. 2.
31. **McCandless, E. L. and Craigie, J. S.**, Sulfated polysaccharides in red and brown algae, *Ann. Rev. Plant Physiol.,* 30, 41, 1979.
32. **Percival, E.**, The polysaccharides of green, red and brown seaweeds: their basic structure, biosynthesis and function, *Br. Phycol. J.,* 14, 103, 1979.
33. **Egami, F. and Takahashi, N.**, Studies on charonin sulfuric acids and sulfatases of *Charonia lampas,* in *Biochemistry and Medicine of Mucopolysaccharides,* Egami, F. and Oshima, Y., Eds., Maruzen, Tokyo, 1962, 53.
34. **Dodgson, K. S. and Rose, F. A.**, Sulfohydrolases, in *Metabolism of Sulfur Compounds,* Greenberg, D. M., Ed. Academic Press, New York, 1975, chap. 9.
35. **Hunt, S.**, *Polysaccharide-Protein Complexes in Invertebrates,* Academic Press, London, 1970.
36. **Katman, R. L. and Jeanloz, R. W.**, Thyonatan 4-sulfate: a sulfated polysaccharide from *Thyone briareus* connective tissue, *J. Biol. Chem.,* 248, 50, 1973.
37. **Vasseur, E.**, *The Chemistry and Physiology of the Jelly Coat of the Sea Urchin Egg,* Kihlstromstryk, Stockholm, 1954.
38. **Ishihara, K., Oguri, K., and Taniguchi, H.**, Isolation and characterization of fucose sulfate from jelly coat glycoprotein of sea urchin egg, *Biochim. Biophys. Acta,* 320, 628, 1973.
39. **Toda, N., Horikawa, T, Anno, K., and Senno, N.**, Galactan sulfate from the test of tunicates, *Carbohydr. Res.,* 62, 389, 1978.
40. **Kwart, H. and Shashoua, V. E.**, The structure and constitution of mucus, *Trans. N.Y. Acad. Sci.,* (Ser. 2), 19, 595, 1957.
41. **Cassaro, C. M. F. and Dietrich, C. P.**, Distribution of sulfated mucopolysaccharides in invertebrates, *J. Biol. Chem.,* 252, 2254, 1977.
42. **Rodén, R.**, Structure and metabolism of connective tissue proteoglycans in *The Biochemistry of Glycoproteins and Proteoglycans,* Lennerz, W. J., Ed. Plenum Press, New York, 1980, chap. 7.
43. **Linker, A.**, Structure of heparan sulfate oligosaccharides and their degradation by exo-enzymes, *Biochem. J.,* 183, 711, 1979.
44. **Cifonelli, J. A. and King, J. A.**, Structural characteristics of heparan sulfates with varying sulfate contents, *Biochemistry,* 16, 2137, 1977.
45. **Doyle, J.**, The "Lorenzan Sulfates": a new group of vertebrate mucopolysaccharides, *Biochem. J.,* 103, 325, 1967.
46. **Lehtonen, A., Kärkkäinen, J., and Haahti, E.**, Carbohydrate components in the epithelial mucin of hagfish, *Myxine glutinosa, Acta Chem. Scand.,* 20, 1456, 1966.

47. **Kancewicz, M.**, The presence of a sulphated polysaccharide in the cell envelopes of *Halobacterium halobium*, *Biochem. J.*, 130, 40p, 1972.
48. **Steber, J. and Schleifer, K. H.**, *Halococcus morrhuae:* A sulfated heteropolysaccharide as the structural component of the bacterial cell wall, *Arch. Microbiol.*, 105, 173, 1975.
49. **McCormick, J. J., Blomquist, J. C., and Rusch, H. P.**, Isolation and characterization of an extracellular polysaccharide from *Physarum polycephalum*, *J. Bacteriol.*, 104, 1110, 1970.
50. **Darby, G. K., Jones, A. S., Kennedy, J. F., and Walker, R. T.**, Isolation and analysis of the nucleic acids and polysaccharides from *Clostridium welchii*, *J. Bacteriol.*, 103, 159, 1970.
51. **Kent, P. W. and Marsden, J. C.**, A sulphated sialoprotein from sheep colonic mucin, *Biochem. J.*, 87, 38p, 1963.
52. **Yosizawa, Z.**, Sulfated glycoproteins, in *Glycoproteins*, (Part B), Gottschalk, A., Ed., Elsevier, Amsterdam, 1972, chap. 8.
53. **Branford-White, C. J. and Appleby, J. A.**, The occurrence of sulphated glycoproteins in nerve endings, *Biochem. Soc. Trans.*, 7, 134, 1979.
54. **Branford-White, C. J. and Mullins, D. J.**, Sulphated glycoproteins in synaptosomal membranes, *Biochem. Soc. Trans.*, 8, 442, 1980.
55. **Mathieu, J. M., Everly, J. L., and Brady, R. O.**, [^{35}S]-Sulfate incorporation into myelin glycoproteins II, peripheral nervous tissue, *Biochim. Biophys. Acta*, 392, 167, 1975.
56. **Embery, G. and Rolla, G.**, Chemistry of the sulfate groups in the sulfated glycoprotein from rabbit submandibular gland, *Scand. J. Dent. Res.*, 85, 241, 1977.
57. **Kato, A., Hirata, S., and Kobayashi, K.**, Structure of the sulfated oligosaccharide chain of ovomucin, *Agric. Biol. Chem.*, 45, 1025, 1978.
58. **Endo, M. and Yosizawa, Z.**, Biosynthesis of sulfated glycoprotein in the endometrium of rabbit uterus, *J. Biochem. Tokyo*, 79, 293, 1976.
59. **Bara, J., Paul-Gardais, A., Loisillier, F., and Burtin, P.**, Isolation of a sulfated glycopeptidic antigen from human gastric tumors: its localization in normal and cancerous gastrointestinal tissues, *Int. J. Cancer*, 21, 133, 1978.
60. **Heifetz, A., Kinsey, W. H., and Lennarz, W. J.**, Synthesis of a novel class of sulfated glycoproteins in embryonic liver and lung, *J. Biol. Chem.*, 255, 4528, 1980.
61. **Ogamo, A., Watanabe, H., and Nagasawa, K.**, Sulfated heteropolysaccharide — peptide complexes from oyster viscera, *J. Biochem. Tokyo*, 76, 573, 1974.
62. **Heifetz, A. and Lennarz, W. J.**, Biosynthesis of *N*-glycosidically linked glycoproteins during gastrulation of sea urchin embryos, *J. Biol. Chem.*, 254, 6119, 1979.
63. **Erickson, J. S. and Kaplan, A. S.**, Synthesis of proteins in cells infected with herpesvirus: IX, sulfated proteins, *Virology*, 55, 94, 1973.
64. **Kaplan, A. S. and Ben-Porat, T.**, Synthesis of proteins in cells infected with herpesvirus: XI, sulfated structural glycoproteins, *Virology*, 70, 561, 1976.
65. **Pinter, A. and Compans, R. W.**, Sulfated components of enveloped viruses, *J. Virol.*, 16, 859, 1975.
66. **Nakamura, K. and Compans, R. W.**, Glycopeptide components of influenza viral glycoproteins, *Virology*, 86, 432, 1978.
67. **Kent, P. W., Coles, C. J., Cooper, J. R., and Mian, N. R.**, Sulfate ester groups as potential informational regulators in glycoproteins, in *Carbohydrate Sulfates*, Schweiger, R. G., Ed., American Chemical Society Symposium Series, American Chemical Society, Washington, D. C., 1978, chap. 3.
68. **Dunham, J. S. and Hynes, R. O.**, Differences in the sulfated macromolecules synthesized by normal and transformed hamster fibroblasts, *Biochim. Biophys. Acta*, 506, 242, 1978.
69. **Marshall, R. D.**, Structures and functions of glycoproteins, *Int. Rev. Biochem.*, 25, 1, 1979.
70. **Soda, T. and Hattori, C.**, An enzyme which hydrolyses glucose mono-sulphate: glucosulphatase, *Bull. Chem. Soc. Jpn.*, 6, 258, 1931.
71. **Dodgson, K. S. and Spencer, B.**, Studies on sulphatases 7: a preliminary account of the glycosulphatase of *Littorina littorea*, *Biochem. J.*, 57, 310, 1954.
72. **Dodgson, K. S.**, Glycosulphatase: observations on the activity of partially purified preparations toward the sulphate esters of certain monosaccharides and steroids, *Biochem. J.*, 78, 324, 1961.
73. **Dodgson, K. S. and Lloyd, A. G.**, Potassium glucose 6-*O*-sulphate as a substrate for glycosulphatase, *Biochem. J.*, 78, 319, 1961.
74. **Dodgson, K. S. and Rose, F. A.**, Sulfohydrolases, in *Metabolism of Sulfur Compounds*, Greenberg, D., Ed., Academic Press, New York, 1975, chap. 9.
75. **Hatanaka, H., Ogawa, Y., and Egami, F.**, Arylsulphatase and glycosulphatase of *Charonia lampas:* substrate specificity toward sugar sulphate derivatives, *Biochem. J.*, 159, 445, 1976.
76. **Tanko, B.**, Splitting of glucose and saccharose sulfuric acids by bacterial sulfatase, *Biochem. Z.*, 247, 486, 1932.
77. **Yamashina, I.**, On the decomposition of glucose sulfate by micro-organisms, *Bull. Chem. Soc. Jpn.*, 72, 124, 1951.

78. **Lloyd, A. G.**, The metabolism of exogenous *N*-acetyl-D-glucosamine 6-0[^{35}S]sulfate in the normal rat, *Biochem. J.*, 80, 572, 1961.
79. **Lloyd, A. G.**, The metabolism of *N*-acetyl-D-galactosamine 6-0[^{35}S]sulphate in the rat, *Biochim. Biophys. Acta*, 58, 1, 1962.
80. **Large, P. J., Lloyd, A. G., and Dodgson, K. S.**, Desulphation of D-glucose 6-sulphate by microorganisms, *Biochem. J.*, 90, 12p, 1964.
81. **Lloyd, A. G., Large, P. J., James, A. M., and Dodgson, K. S.**, The metabolic fate of hexose 6-0-[^{35}S]sulphates and related compounds, *J. Biochem. Tokyo*, 55, 669, 1964.
82. **Lloyd, A. G., Large, P. J., Davies, M., Olavesen, A. H., and Dodgson, K. S.**, The glycosulphatase of *Trichoderma viride*, *Biochem. J.*, 108, 393, 1968.
83. **Dodgson, K. S.**, Determination of inorganic sulfate in studies on the enzymic and non-enzymic hydrolysis of carbohydrate and other sulfate esters, *Biochem. J.*, 78, 312, 1961.
84. **Dodgson, K. S., Lloyd, A. G., and Tudball, N.**, *O*-Sulfate esters of L-serine, L-threonine and L-hydroxyproline, *Biochem. J.*, 79, 111, 1961.
85. **Weigl, J. and Yaphe, W.**, The enzymic hydrolysis of carrageenan by *Pseudomonas carrageenovora:* purification of ϰ-carrageenase, *Can. J. Microbiol.*, 12, 939, 1966.
86. **Weigl, J. and Yaphe, W.**, Glycosulfatase of *Pseudomonas carrageenovora:* desulfation of disaccharide from ϰ-carrageenan, *Can. J. Microbiol.*, 12, 874, 1966.
87. **McLean, M. W. and Williamson, F. B.**, Glycosulfatase from *Pseudomonas carrageenovora:* purification and some properties, *Eur. J. Biochem.*, 101, 497, 1979.
88. **Nakazawa, K., Suzuki, N., and Suzuki, S.**, Sequential degradation of keratan sulfate by bacterial enzyme and purification of a sulfatase in the enzymatic system, *J. Biol. Chem.*, 250, 905, 1975.
89. **Nakazawa, K., and Kagabe, K.**, Galactose 6-sulfatase from *Actinobacillus* sp. IF0-13310 and its action on sulfated oligosaccharides from keratan sulfate, *Biochim. Biophys. Acta*, 527, 391, 1978.
90. **Fitzgerald, J. W. and Dodgson, K. S.**, Sulphur utilization during growth of *Pseudomonas fluorescens* on potassium D-glucose 6-*O*-sulphate, *Biochem. J.*, 121, 521, 1971.
91. **Fitzgerald, J. W. and Dodgson, K. S.**, Carbon and sulphur utilization during growth of *Pseudomonas fluorescens* on potassium D-glucose 6-*O*-sulphate as the sole sulphur source, *Biochem. J.*, 122, 277, 1971.
92. **Lloyd, A. G., Large, P. J., Tudball, N., and Dodgson, K. S.**, Kinetic studies on the oxidation of D-glucose 6-*O*-sulphate by glucose 6-phosphate dehydrogenase, *Biochem. J.*, 97, 42P, 1965.
93. **Hauge, J. G.**, Glucose dehydrogenases-particulate. I. *Pseudomonas* species and *Bacterium anitratum*, *Methods Enzymol.*, 9, 92, 1966.
94. **Evans, R. W., Kushwaha, S. C., and Kates, M.**, The lipids of *Halobacterium marismortui*, an extremely halophilic bacterium in the Dead Sea, *Biochim. Biophys. Acta*, 619, 533, 1980.
95. **Araki, B. and Arai, K.**, Chemical constitution of agar/agar, Part 18: isolation of a new crystalline disaccharide by enzymic hydrolysis of agar/agar, *Bull. Chem. Soc. Jpn.*, 29, 339, 1956.
96. **Yaphe, W. and Baxter, B.**, The enzymic hydrolysis of carrageenan, *Appl. Microbiol.*, 3, 380, 1955.
97. **Rees, D. A.**, Structure, conformation and mechanism in the formation of polysaccharide gels and networks, *Adv. Carbohydr. Chem.*, 24, 267, 1969.
98. **McLean, M. W., and Williamson, F. B.**, ϰ-Carrageenase from *Pseudomonas carrageenovora*, *Eur. J. Biochem.*, 93, 553, 1979.
99. **Johnston, K. H. and McCandless, E. L.**, Enzymic hydrolysis of the potassium chloride soluble fraction of carrageenan: properties of "λ-carrageenases" from *Pseudomonas carrageenovora*, *Can. J. Microbiol.*, 19, 779, 1973.
100. **Neuberg, C. and Rubin, O.**, Uber die Bildung von Thioschwefelsäure aus Atherschioefelsäuren und Sulfonsäuren, *Biochem. Z.*, 67, 82, 1914.
101. **Neuberg, C. and Hofmann, E.**, Enzymatischer abbau von Chondroitin Schwefelsaure, *Biochem. Z.*, 234, 345, 1931.
102. **Candeli, A. and Tronieri, A.**, Bacterial sulphatases and dental caries, *Boll. Soc. Ital. Biol. Sper.*, 27, 651, 1951.
103. **Regianini, O.**, Enzymes hydrolysing chondroitin sulphuric acid, *Boll. Soc. Ital. Biol. Sper.*, 26, 866, 1950.
104. **Hartles, R. L. and McLean, N. D.**, Some observations on the biochemical properties of a strain of *Clostridium welchii* isolated from carious teeth, *Br. Dent. J.*, 93, 147, 1952.
105. **Pincus, P.**, A sulphated mucopolysaccharide in human dentine, *Nature (London)*, 166, 187, 1950.
106. **Pincus, P.**, Production of dental caries: a new hypothesis, *Br. Med. J.*, 2, 358, 1949.
107. **Schultz-Haudt, S. D. and Scherp, H. W.**, The production of chondrosulphatase by microorganisms isolated from human gingival crevices, *J. Dent. Res.*, 35, 299, 1956.
108. **Smith, R. F. and Willett, N. P.**, Rapid plate method for screening hyaluronidase and chondroitin sulfatase — producing microorganisms, *Appl. Microbiol.*, 16, 1434, 1968.

109. **Hershon, L. E.**, Elaboration of hyaluronidase and chondroitin sulfatase by microorganisms inhabiting the gingival sulcus: evaluation of a screening method for periodontal disease, *J. Peridont.*, 42, 34, 1971.
110. **Lederman, H. M., Hayashi, J. A., and Bahn, A. N.**, Chondroitin sulfatases in oral bacteria, *J. Dent. Res.*, 51, 206, 1972.
111. **Dodgson, K. S., Lloyd, A. G., and Spencer, B.**, Studies on sulphatases, 14: a preliminary account of the chondrosulphatase of *Proteus vulgaris*, *Biochem. J.*, 65, 131, 1957.
112. **Dodgson, K. S. and Lloyd, A. G.**, Studies on sulphatases, 18: preparation of chondroitinase-free chondrosulphatase from extracts of *Proteus vulgaris*, *Biochem. J.*, 66, 532, 1957.
113. **Konelza, W. A., Pelczar, M. J., and Burnett, G. W.**, The degradation of chondroitin sulfate by resting cells of *Proteus vulgaris*, *Bact. Proc.*, (Abstr. P37), 106, 1954.
114. **Dodgson, K. S. and Lloyd, A. G.**, The degradation of chondroitin sulphate by the chondroitinase of *Proteus vulgaris*, *Biochem. J.*, 68, 88, 1958.
115. **Martinez, R. J., Wolfe, J. B., and Nakada, H. J.**, Purification and properties of the enzyme chondroitinase, *J. Biol. Chem.*, 234, 2236, 1959.
116. **Linker, A., Hoffman, P., Meyer, K., Sampson, P., and Korn, E. D.** The formation of unsaturated disaccharides from mucopolysaccharides and their cleavage to α-keto acid by bacterial enzymes, *J. Biol. Chem.*, 235, 3061, 1960.
117. **Suzuki, S.**, Isolation of novel disaccharides from chondroitin sulfates, *J. Biol. Chem.*, 235, 3580, 1960.
118. **Nakada, H. I. and Wolfe, J. B.**, Studies on the enzyme chondroitinase: product structure and ion effects, *Arch. Biochem. Biophys.*, 94, 2441, 1961.
119. **Yamagata, T., Kawamura, Y., and Suzuki, S.**, Substrate specificity of two separate chondrosulfatases in *Proteus vulgaris*, *Biochim. Biophys. Acta*, 115, 250, 1966.
120. **Yamagata, T., Saito, H., Habuchi, O., and Suzuki, S.**, Purification and properties of bacterial chondroitinases and chondrosulfatases, *J. Biol. Chem.*, 243, 1523, 1968.
121. **Suzuki, S., Saito, H., Yamagata, T., Anno, K., Seno, N., Kawai, Y., and Furuhashi, T.**, Formation of three types of disulfated disaccharides from chondroitin sulfates by chondroitinase digestion, *J. Biol. Chem.*, 243, 1543, 1968.
122. **Thurston, C. F., Hardingham, T. E., and Muir, H.**, The kinetics of degradation of chondroitin sulphates and hyaluronic acid by chondroitinase from *Proteus vulgaris*, *Biochem. J.*, 145, 397, 1975.
123. **Seno, N., Akiyama, F., and Anno, K.**, Substrate specificity of chondrosulfatases from *Proteus vulgaris* for sulfated tetrasaccharides, *Biochim. Biophys. Acta*, 362, 290, 1974.
124. **Payza, A. N. and Korn, E. D.**, The degradation of heparin by bacterial enzymes, 1: adaptation and lyophilized cells, *J. Biol. Chem.*, 223, 853, 1956.
125. **Christensen, P.**, Description and taxonomic status of *Cytophaga heparina* (Payza and Korn) comb. nov. (basionym: *Flavobacterium heparinum* Payza and Korn 1956), *Int. J. Syst. Bacteriol.*, 30, 473, 1980.
126. **Hoffman, P., Linker, A., Sampson, P., Meyer, K., and Korn, E. D.**, The degradation of hyaluronate, the chondroitin sulfates and heparin by bacterial enzymes *(Flavobacterium)*, *Biochim. Biophys. Acta*, 25, 658, 1957.
127. **Michelacci, Y. M. and Dietrich, C. P.**, Studies on the induction of a chondroitinase in *Flavobacterium heparinum*, *Biochimie*, 55, 893, 1973.
128. **Michelacci, Y. M. and Dietrich, C. P.**, Isolation and partial characterization of an induced chondroitinase B from *Flavobacterium heparinum*, *Biochem. Biophys. Res. Comm.*, 56, 973, 1974.
129. **Michelacci, Y. M. and Dietrich, C. P.**, Chondroitinase C from *Flavobacterium heparinum*, *J. Biol. Chem.*, 251, 1154, 1976.
130. **Michelacci, Y. M. and Dietrich, C. P.**, A comparative study between a chondroitinase B and a chondroitinase AC from *Flavobacterium heparinum*, *Biochem. J.*, 151, 121, 1975.
131. **Otatani, N. and Yosizawa, Z.**, Purification of chondroitinase B and chondroitinase C using glycosaminoglycan-bound AH-Sepharose® 4B, *Carbohydr. Res.*, 70, 295, 1979.
132. **Salyers, A. A. and O'Brien, M.**, Cellular location of enzymes involved in chondroitin sulfate breakdown by *Bacteroides thetaiotaomicron*, *J. Bacteriol.*, 143, 772, 1980.
133. **Salyers, A. A. and Kotarski, S. F.**, Induction of chondroitin sulfate lyase activity in *Bacteroides thetaiotaomicron*, *J. Bacteriol.*, 143, 781, 1980.
134. **Levinson, M. E.**, The importance of anaerobic bacteria in infectious diseases, *Med. Clin. N. Am.*, 57, 1015, 1973.
135. **Willis, A. T., Young, S. E. J., and Ferguson, I. R.**, Infections due to non-sporing anaerobes, some illustrative cases, in *Infection with Non-Sporing Anaerobic Bacteria*, Philips, I. and Sussman, M., Eds., Churchill Livingstone, Edinburgh, 1974, 189.
136. **Hoeffler, U.**, Enzymatic and hemolytic properties of *Propionibacterium acnes* and related bacteria, *J. Clin. Microbiol.*, 6, 555, 1977.

137. **Rudek, W. and Haque, R-V.**, Extracellular enzymes of the genus *Bacteroides*, *J. Clin. Microbiol.*, 4, 458, 1976.
138. **Rudek, W. and Haque, R-V.**, Purification of a mucopolysaccharidase from *Bacteroides distasonis*, *J. Gen. Microbiol.*, 119, 211, 1980.
139. **Steffen, E. K., Hentges, D. J., and Campbell, B. J.**, Selected enzymes of clinical anaerobic isolates, *Abstr. Annu. Meet. Am. Soc. Microbiol.*, (Abstr. B22), 19, 1979.
140. **Helderman, V. P. and Hoogeveen, C. J. C. M.**, Bacterial enzymes and viable counts in crevices of non-inflamed and inflamed gingiva, *J. Periodont. Res.*, 11, 25, 1976.
141. **Kendell, K. and Elvin-Lewis, M.**, Chondroitin sulfatase and hyaluronidase activity of endodontic streptococcal isolates, *J. Dent. Res.*, 52, 132, 1973.
142. **Kitamikado, M. and Lee, Y-Z.**, Chondroitinase-producing bacteria in natural habitats, *Appl. Microbiol.*, 29, 414, 1975.
143. **Hiyama, K. and Okada, S.**, Crystallization and some properties of chondroitinase from *Arthrobacter aurescens*, *J. Biol. Chem.*, 250, 1824, 1975.
144. **Hiyama, K. and Okada, S.**, Amino acid composition and physiocochemical characterization of chondroitinase from *Arthrobacter aurescens*, *J. Biochem. Tokyo*, 78, 1183, 1975.
145. **Hiyama, K. and Okada, S.**, The mode of action of two chondroitinase-AC preparations of different origin, *J. Biochem. Tokyo*, 80, 1201, 1976.
146. **Hiyama, K.**, Action of chondroitinases II. Numerical calculation of the degree of multiple attack, *J. Biochem. Tokyo*, 80, 1209, 1976.
147. **Hiyama, K. and Okada, S.**, Multiple forms of chondroitinase AC from *Arthrobacter aurescens*, *Agric. Biol. Chem.*, 41, 1279, 1977.
148. **Makarem, E. H. and Berk, R. S.**, Partial purification and characterization of chondroitinase from *Proteus mirabilis*, *J. Infect. Dis.*, 118, 427, 1968.
149. **Saito, H., Yamagata, T., and Suzuki, S.**, Enzymatic methods for the determination of small quantities of isomeric chondroitin sulfates, *J. Biol. Chem.*, 243, 1536, 1968.
150. **Murata, K., Harada, T., and Okubo, K.**, Enzymatic studies of chondroitin sulfates in human arterial tissue, *J. Atheroscler. Res.*, 8, 951, 1968.
151. **Murao, P. A. S., Toledo, S. P. A., Nader, H. B., and Dietrich, C. P.**, Excretion of chondroitin sulfate C with low sulfate content by patients with generalized platyspondyly (brachyolmia), *Biochem. Med.*, 7, 415, 1973.
152. **Takeuchi, J.**, Effect of chondroitinases on the growth of solid Ehrlich ascites tumour, *J. Cancer*, 26, 115, 1972.
153. **Rosen, O., Hoffman, P., and Meyer, K.**, Enzymatic hydrolysis of keratosulfate, *Fed. Proc. Fed. Am. Soc. Exp. Biol.*, 19, 147, 1960.
154. **Hirano, S. and Meyer, K.**, Enzymatic degradation of corneal and cartilaginous keratosulfates, *Biochem. Biophys. Res. Comm.*, 44, 1371, 1971.
155. **Hirano, S. and Meyer, K.**, Purification and properties of a keratan sulfate hydrolyzing enzyme, an endo-β-galactosidase, *Connect. Tiss. Res.*, 2, 1, 1973.
156. **Kitamikado, M., Ueno, R., and Nakamura, T.**, Enzymatic degradation of whale cartilage keratosulfate: I, isolation of a keratosulfate-degrading bacterium, *Bull. Jpn. Soc. Sci. Fish.*, 36, 592, 1970.
157. **Kitamikado, M., Ueno, R., and Nakamura, T.**, Enzymatic degradation of whale cartilage keratosulfate: II, identification of a keratosulfate-degrading bacterium, *Bull. Jpn. Soc. Sci. Fish.*, 36, 1172, 1970.
158. **Kitamikado, M. and Ueno, R.**, Enzymatic degradation of whale cartilage keratosulfate: III, purification of a bacterial keratosulfate-degrading enzyme *Bull. Jpn. Soc. Sci. Fish.*, 36, 1175, 1970.
159. **Fukuda, M. N. and Matsumura, G.**, Endo-β-galactosidase of *Escherichia freundii*, hydrolysis of pig colonic mucin and milk oligosaccharides by endoglycosidic action, *Biochem. Biophys. Res. Comm.*, 64, 465, 1975.
160. **Fukuda, M. N. and Matsumura, G.**, Endo-β-galactosidase of *Escherichia freundii:* purification and endoglycosidic action on keratan sulfates, oligosaccharides and blood group active glycoprotein, *J. Biol. Chem.*, 251, 6218, 1976.
161. **Nakazawa, K. and Suzuki, S.**, Purification of keratan sulfate-endogalactosidase and its action on keratan sulfates of different origin, *J. Biol. Chem.*, 250, 912, 1975.
162. **Lindahl, V.**, Structure and biosynthesis of L-iduronic acid-containing glycosaminoglycans, in *MTP International Review of Science: Organic Chemistry Series Two-Carbohydrate Chemistry*, Aspinall, G. O., Ed., Butterworths, London, 1976, 283.
163. **Lindahl, V. and Höök, M.**, Glycosaminoglycans and their binding to biological macromolecules, *Annu. Rev. Biochem.*, 47, 385, 1978.
164. **Korn, E. D. and Payza, A. N.**, Enzymic degradation of heparin, *Biochim. Biophys. Acta*, 20, 596, 1956.
165. **Korn, E. D. and Payza, A. N.**, The degradation of heparin by bacterial enzymes: II acetone powder extracts, *J. Biol. Chem.*, 223, 859, 1956.

166. **Korn, E. D.**, The degradation of heparin by bacterial enzymes: III, a comparison of the degradation of heparin, hyaluronic acid, and chondroitin sulfate, *J. Biol. Chem.*, 226, 841, 1957.
167. **Linker, A. and Hovingh, P.**, The enzymatic degradation of heparin and heparitin sulfate: 1, the fractionation of a crude heparinase from flavobacteria, *J. Biol. Chem.*, 240, 3724, 1965.
168. **Dietrich, C. P.**, Novel heparin degradation products: isolation and characterization of novel disaccharides and oligosaccharides produced from heparin by bacterial degradation, *Biochem. J.*, 108, 647, 1968.
169. **Dietrich, C. P.**, Enzymic degradation of heparin: a sulfamidase and a sulfoesterase from *Flavobacterium heparinum, Biochem. J.*, 111, 91, 1969.
170. **Dietrich, C. P.**, Enzymic degradation of heparin: a glucosaminidase and a glycuronidase from *Flavobacterium heparinum, Biochemistry*, 8, 2089, 1969.
171. **Karapally, J. C. and Dietrich, C. P.**, A uronic acid isomerase in *Flavobacterium heparinum, Can. J. Biochem.*, 48, 164, 1970.
172. **Hovingh, P. and Linker, A.**, The enzymatic degradation of heparin and heparitin sulfate: III, purification of a heparitinase and a heparinase from flavobacteria, *J. Biol. Chem.*, 245, 6170, 1970.
173. **Perlin, A. S., Mackie, D. M., and Dietrich, C. P.**, Evidence for a (1→4)-linked 4-*O*(α-L-idopyronosyluronic acid 2-sulfate)-(2-deoxy-2-sulfoamino-D-glucopyranosyl 6-sulfate) sequence in heparin, *Carbohydr. Res.*, 18, 185, 1971.
174. **Linker, A. and Hovingh, P.**, Isolation and characterization of oligosaccharides obtained from heparin by the action of heparinase, *Biochemistry*, 11, 563, 1972.
175. **Warnick, C. T. and Linker, A.**, Purification of an unusual α-glucuronidase from flavobacteria, *Biochemistry*, 11, 568, 1972.
176. **Dietrich, C. P., Silva, M. E., and Michelacci, Y. M.**, Sequential degradation of heparin in *Flavobacterium heparinum* purification and properties of five enzymes involved in heparin degradation, *J. Biol. Chem.*, 248, 6408, 1973.
177. **Lloyd, A. G., Fowler, L. J., Embery, G., and Law, B. A.**, Degradation of [^{35}S] heparin by mammalian and bacterial sulfamidases, *Biochem. J.*, 110, 54p, 1968.
178. **Linker, A., Hoffman, P., Sampson, P., and Meyer, K.**, Heparitin sulfate, *Biochim. Biophys. Acta*, 29, 443, 1958.
179. **Linker, A. and Sampson, P.**, The enzymatic degradation of heparitin sulfate, *Biochim. Biophys. Acta*, 43, 366, 1960.
180. **Linker, A. and Hovingh, P.**, The enzymatic degradation of heparitin sulfate: II, isolation and characterization of non-sulfated oligosaccharides, *Biochim. Biophys. Acta*, 165, 89, 1968.
181. **Hovingh, P. and Linker, A.**, The disaccharide repeating units of heparan sulfate, *Carbohydr. Res.*, 37, 181, 1974.
182. **Linker, A. and Hovingh, P.**, The heparitin sulfates (heparan sulfates), *Carbohydr. Res.*, 29, 41, 1973.
183. **Silva, M. E. and Dietrich, C. P.**, Isolation and partial characterization of three induced enzymes from *Flavobacterium heparinum* involved in the degradation of heparin and heparitin sulfates, *Biochem. Biophys. Res. Comm.*, 56, 965, 1974.
184. **Silva, M. E., Dietrich, C. P., and Nader, H. B.**, On the structure of heparitin sulfates: analyses of the products formed from heparitin sulfates by two heparitinases and a heparinase from *Flavobacterium heparinum, Biochim. Biophys. Acta*, 437, 129, 1976.
185. **Silva, M. E. and Dietrich, C. P.**, Structure of heparin: characterization of the products formed from heparin by the action of a heparinase and a heparitinase from *Flavobacterium heparinum, J. Biol. Chem.*, 250, 6841, 1975.
186. **Lloyd, A. G., Law, B. A., Fowler, L. J., and Embery, G.**, Bacterial carbohydrate sulfamidase induction and heparin degradation, *Biochem. J.*, 110, 54p, 1968.
187. **Dietrich, C. P.**, Studies on the induction of heparin degrading enzymes in *Flavobacterium heparinum, Biochemistry*, 8, 3342, 1969.
188. **Silva, M. E. and Dietrich, C. P.**, Studies on the induction of heparin degrading enzymes in *Flavobacterium heparinum*, II. Structural requirements of the inducer, *Biochimie*, 55, 1101, 1973.
189. **Gesner, B. M. and Jenkin, C. R.**, Production of heparinase by *Bacteroides, J. Bacteriol.*, 81, 595, 1961.
190. **Hovingh, P. and Linker, A.**, Specificity of flavobacterial glycuronidases acting on disaccharides derived from glycosaminoglycans, *Biochem. J.*, 165, 287, 1977.
191. **Dietrich, C. P. and Dietrich, S. M. C.**, Simple micromethod for identification of heparin and other acidic mucopolysaccharides from mammalian tissues, *Analyt. Biochem.*, 46, 209, 1972.
192. **Linker, A. and Hovingh, P.**, The uses of degradative enzymes as tools for identification and structural analysis of glycosaminoglycans, *Fed. Proc. Fed. Am. Soc. Exp. Biol.*, 36, 43, 1977.
193. **Farooqui, A. A.**, Sulfatases, sulfate esters and their metabolic disorders, *Clin. Chim. Acta*, 100, 285, 1980.

Chapter 4

ARYLSULFATASES

I. INTRODUCTION

Arylsulfatases (arylsulfate sulfohydrolases, EC 3.1.6.1) catalyze the hydrolysis of aryl sulfates according to Reaction 1, and occur very widely distributed in the animal, plant, and microbial worlds.

$$ArO{-}SO_3^- + H_2O \rightarrow ArOH + SO_4^{2-} + H^+ \qquad (1)$$

The substrates most commonly used for their assay are simple chromogenic arylsulfates such as *p*-nitrophenyl sulfate or 2-hydroxy-5-nitrophenyl sulfate (nitrocatechol sulfate or 4-nitrocatechol sulfate). The activity of arylsulfatases toward such compounds is relatively easy to assay and this probably accounts at least in part for the large volume of literature concerning these enzymes that has accumulated to date. Current interest in these enzymes is high, not least because congenital defects leading to an absence or deficiency of certain of the mammalian arylsulfatases are characterized by certain disease states in man.[1,2] As will emerge in the following paragraphs, mammals protect themselves against the toxicity of phenolic compounds by conjugation with, among other moieties, the sulfate group, and it therefore seemed odd that enzymes should also be present to reverse the process. However, it is now well established[3] that the mammalian arylsulfatases possess activity toward other naturally occurring sulfate esters, and it is the accumulation of these compounds in the absence of effective degrading enzymes that leads to inherited metabolic disorders. In short, the mammalian enzymes are not concerned in vivo with the hydrolysis of the simple aryl sulfates which were instrumental in their discovery.

Anticipating in the most general terms some of what follows in Volume II, Chapters 3 and 4, microbial arylsulfatases are considered[4] to participate in two types of metabolism, namely assimilation of inorganic sulfate and dissimilation of aryl compounds. In the first case it is understandable that when an organism encounters a medium lacking inorganic sulfate, cysteine, or intervening intermediates in the assimilation pathway, it must resort to using other less directly assimilatable sulfur sources such as aryl sulfate esters. In the second case, aryl compounds derived from aryl sulfates may be required as a carbon source when other appropriate carbon compounds are absent. In either of these situations separation of sulfate from the aryl nucleus by an arylsulfatase would be a logical first step. Such a role for microbial arylsulfatases immediately raises the question as to the types of aryl sulfate esters that microorganisms are likely to encounter in soils, rivers, and the oceans. This is an important consideration because it puts the microbial arylsulfatases into a perspective in terms of available natural substrates.

II. NATURALLY OCCURRING ARYL SULFATE ESTERS

A. Simple Phenyl Sulfates

The ability of a wide range of creatures to conjugate toxic phenols with inorganic sulfate is well documented.[5-7] Over a century ago Eugene Baumann[8] first discovered that dogs fed a diet containing phenol produced phenyl sulfate (Figure 1) in their urine. Since that time, reports of sulfate conjugation of phenolic compounds have become legion. Phenols are not only produced by microorganisms in the gastrointestinal tract,[9-12] but are also present in food (both as natural constituents and as additives)

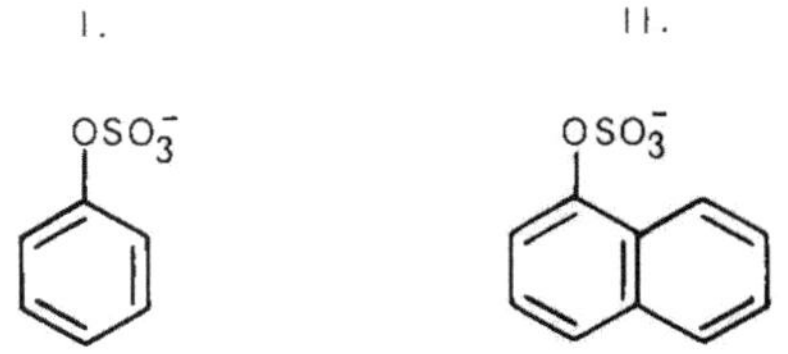

FIGURE 1. Simple naturally occurring aryl sulfates. I, phenyl sulfate; II, naphthyl 1-sulfate.

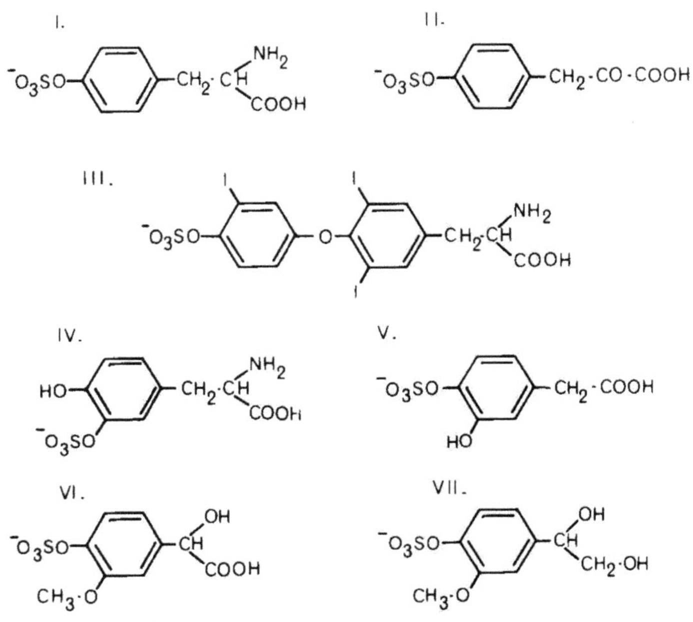

FIGURE 2. Naturally occurring aryl sulfates derived from tyrosine. I, tyrosine-*O*-sulfate; II, 4-hydroxyphenylpyruvic acid 4-sulfate; III, 3:5:3′-triiodothyronine 4′-sulfate; IV, dopamine 3-*O*-sulfate; V, 3,4-dihydroxyphenylacetic acid 4-sulfate; VI, 3-methoxy 4-hydroxymandelic acid 4-sulfate; VII, 3-methoxy 4-hydroxyphenylglycol 4-sulfate.

and in the environment to which various species are exposed. For example, the oil-based technology which is currently enjoyed has increased exposure of many living creatures to a wide range of aromatic compounds. Thus aryl sulfates (e.g., naphthyl 1-sulfate, Figure 1) are produced by marine organisms[13,14] during exposure to aromatic hydrocarbons and related compounds, a situation which might well occur following for example, a crude oil spillage at sea.[15] Terrestrial animals cope in a similar way[16-21] with aryl hydrocarbons and phenols that are present, for example, in coal-tar products.

B. Tyrosine-*O*-Sulfate

The *O*-sulfate of the amino acid tyrosine (Figure 2,I) was first discovered in a peptide from fibrinogen.[22] It has since been found in other fibrinopeptides isolated from several mammals,[23] and in gastrin II[24] (a stomach hormone in the pig), cholecystokinin-pancreozymin[25] (a pig pancreas hormone), phyllokinin,[26] and caerulein[27] (peptides in the skin of amphibians). Tyrosine*O*-sulfate and its deaminated analogue 4-hydroxyphenylpyruvic acid sulfate (Figure 2,II) also occur in the free state in mammalian urine[28-31] and in some marine invertebrates.[32]

C. Thyroid Hormone Sulfates

3:5:3′-Triiodothyronine-*O*-sulfate (Figure 2,III) has been found in rat bile and plasma,[33] and beef thyroid.[34] Human liver extracts[35] contain enzymes capable of sulfoconjugation of the parent thyroid hormones.

D. Catecholamine-*O*-Sulfates

Dopamine 3-*O*- and 4-*O*-sulfates (Figure 2,IV) occur in urine and plasma of healthy humans[36-38] and of patients[39,40] with Parkinson's disease. These conjugates have also been found in tissues of various other creatures.[41-43] The related compounds 3,4-dihydroxyphenylacetic acid 4-sulfate (Figure 2,V), 3-methoxy 4-hydroxyphenylacetic acid 4-sulfate, and 3-methoxy 4-hydroxymandelic acid 4-sulfate (Figure 2,VI) have been detected in human urine,[44] and the occurrence of 3-methoxy 4-hydroxyphenylglycol 4-sulfate (Figure 2,VII) in nervous tissues is well established.[45-49]

E. Vitamin B_6 Derivatives

Following the indications that excess vitamin B_6 was excreted in human and dog urine as a sulfate conjugate,[50] Sata and his colleagues[51] showed that rat liver preparations were capable of conjugating pyridoxine and pyridoxal to the corresponding 3-*O*-sulfates (Figure 3,I).

F. Aryl Sulfates Based on the Indole Nucleus

The reader will recall from Volume I, Chapter 1 the interesting history and biochemistry of the Tyrian purple dyes,[52] the precursors for which are indoxyl- and 6-bromoindoxyl 3-*O*-sulfates (Figure 3,II and III) from marine mollusks.[53-56] Other naturally occurring aryl sulfate esters based on the indole nucleus are serotonin sulfate (Figure 3,IV) in urine[57,58] and nervous tissue;[43] 5-hydroxyindoxyl acetic acid sulfate in urine;[58] indoxyl 3-*O*-sulfate in urine of several mammals;[29] 5-, 6-, and 7-hydroxyskatole-*O*-sulfates[59] (Figure 3, V) and 2-amino-3-hydroxyacetophenone 3-*O*-sulfate[60](Figure 3,VI) in human urine; and bufothionine[61,62] (Figure 3,VII) and the related bufoviridine[63] (dihydrobufothionine) and bufoviridine-*N*-sulfonate[64] in the skin and secretions of toads. In addition, ommatin D (Figure 3,VIII) is an aryl sulfate pigment widely found in the excretions and wings of insects[65,66] and although not itself an indole structure, it is probably derived from tryptophan or a related compound.

G. Steroid Sulfates

Estrone 3-sulfate (Figure 4,I) occurs in the urine of pregnant women[67] and mares,[68] and is also an important circulating estrogen.[69] Sulfate esters of related compounds also occur naturally such as estradiol 3-sulfate (Figure 4,II) in sea urchins[70] and in human cord blood and other fluids,[71] 16-hydroxyestrone 3-sulfate (Figure 4,III) in human urine,[72] and estriol 3-sulfate (Figure 4,IV) in blood during pregnancy[73] and in the meconium of the fetus.[74] In addition several tissues have been shown to possess the ability to transfer sulfate to steroids, e.g., adrenal glands[75,76] and larval gut tissue.[77]

H. Plant Flavonoid Sulfates

It is now more than 40 years[78] since the first discovery of a flavonoid sulfate, isorhamnetin 3-sulfate (persicarin, Figure 5,I) in the water pepper *Polygonum hydropiper.* By the beginning of the 1970s, only a few more flavonoid sulfates had been isolated and it is hardly surprising that they were considered among the rarest of naturally occurring flavonoids. However, during the past decade there has been an enormous increase in the rate of discovery of flavonoid sulfate esters in plants; so much so that they are currently regarded as an important class of natural compounds.[79,80] Over 50 different flavonoid sulfates have been detected in surveys of 200 species drawn from

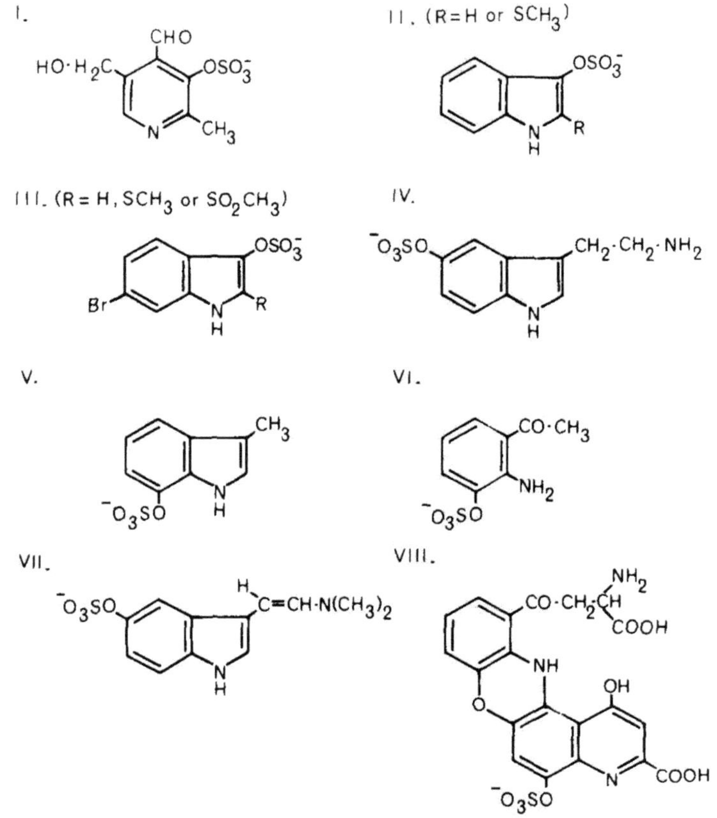

FIGURE 3. Naturally occurring aryl sulfates derived from nitrogenous heterocyclic compounds. I, pyridoxal 3-sulfate; II, indoxyl 3-sulfate; III, 6-bromoindoxyl 3-sulfate; IV, serotonin sulfate; V, skatole 7-sulfate; VI, 2-amino 3-hydroxyacetophenone 3-sulfate; VII, bufothionine; VIII, ommatin D.

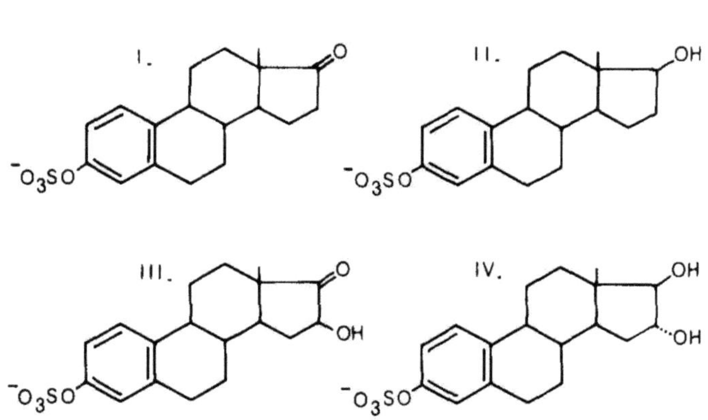

FIGURE 4. Naturally occurring estrogen aryl sulfates. I, estrone 3-sulfate; II, estradiol 3-sulfate; III, 16-hydroxyestrone 3-sulfate; IV, estriol 3-sulfate.

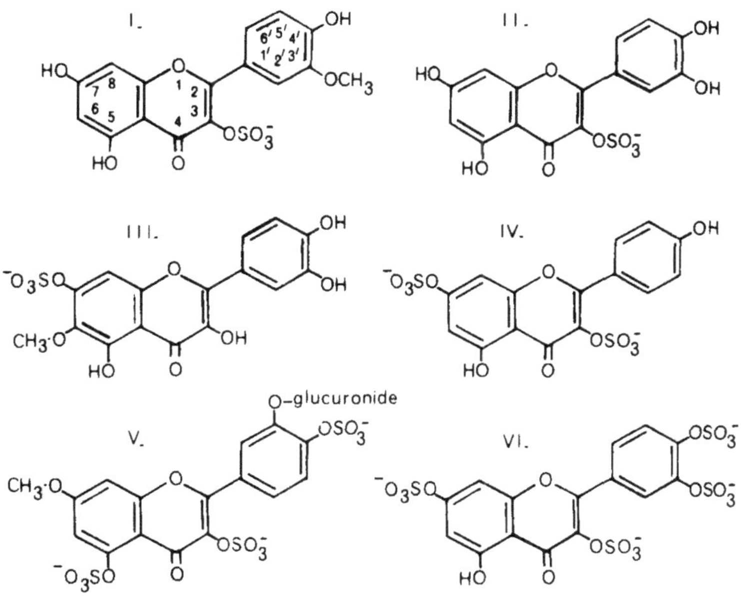

FIGURE 5. Naturally occurring flavonol sulfates. I, isorhamnetin 3-sulfate; II, quercetin 3-sulfate; III, patuletin 7-sulfate; IV, kaempferol 3,7-disulfate; V, rhamnetin 3,5,4′-trisulfate 3′-glucuronide; VI, quercetin, 3,7,3′,4′-tetrasulfate.

20 plant families, particularly higher plants which prefer marine, estuarine, or fresh water environments.[81-83] Although some of these compounds contain sugar residues as well, the majority have one or more sulfate groups attached to phenolic hydroxyl groups. Some examples will serve to illustrate the types which occur, but it must be remembered that they are representative of large numbers of similar compounds already known and that many more are likely to be discovered.

The two most common flavonol monosulfates are isorhamnetin 3-sulfate (Figure 5,I) and the corresponding quercetin analogue (Figure 5,II) which co-occur in several plants.[79] Strictly speaking the 3-hydroxy group of the flavonols is not a true phenolic OH group, but the flavonol structure is probably sufficiently aromatic for its 3-sulfate to be considered as an aryl ester. True aryl sulfate esters also occur in flavonols such as isorhamnetin 7-sulfate,[84] quercetin 7-sulfate 3-glucuronide,[84] patuletin 7-sulfate (Figure 5,III), and patuletin 7-sulfate 3-glucuronide.[85]

In recent years, several flavonol polysulfates have been discovered such as quercetin 3,4′-disulfate,[86] isorhamnetin 3,7-disulfate,[87] and kaempferol 3,7-disulfate[88] (Figure 5,IV). Rhamnetin forms a 3,5,4′-trisulfate 3′-glucuronide[89] (Figure 5,V) which is also notable as a rare example of sulfation at position 5. Quercetin with five phenolic hydroxyl groups occurs as the 3,7,4′-trisulfate,[86] the 3-acetyl 3′,4′,7-trisulfate,[90] and the 3,7,3′, 4′-tetrasulfate[91] (Figure 5,VI).

Most flavones (i.e., lacking the 3-OH group) are based on apigenin, luteolin, or related compounds. The simplest and most common are the 7-sulfates, and such derivatives of apigenin[84] (Figure 6,I), luteolin[84] (Figure 6,II), diosmetin[92] (luteolin 4′-methyl ether), chrysoeriol[81] (luteolin 3′-methyl ether), and tricetin[92] (Figure 6,III) are all known. Sulfation at other hydroxyls also occurs as in luteolin 4′-sulfate[79] and hypolaetin 8-sulfate[84] (Figure 6,IV). The 3′,7′-disulfates of luteolin, diosmetin, and tricetin have also been found.[81,92] The family Polygonaceae that are among these plants producing flavonoid sulfates also yield some other phenolic sulfates including what appear to be 3-sulfates of caffeic and gallic acids (Figure 7,I and II).[83]

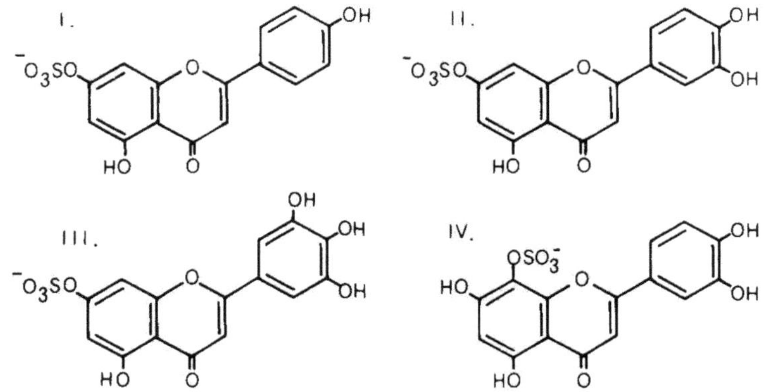

FIGURE 6. Naturally occurring flavone sulfates. I, apigenin 7-sulfate; II, luteolin 7-sulfate; III, tricetin 7-sulfate; IV, hypolaetin 8-sulfate.

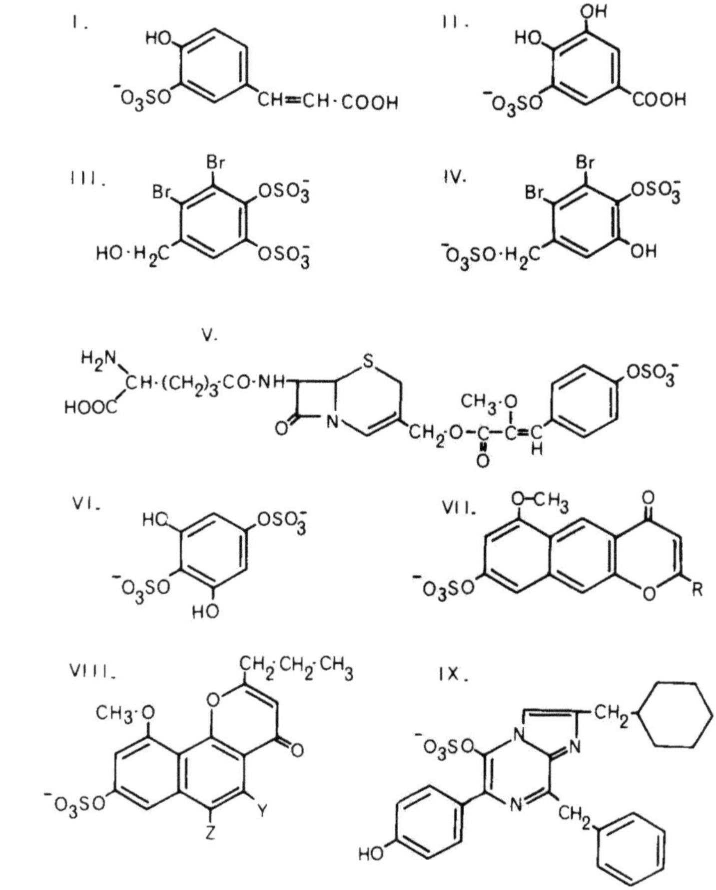

FIGURE 7. Other naturally occurring aryl sulfates. I, caffeic acid 3-sulfate; II, gallic acid 3-sulfate; III, 2,3-dibromo-4,5-dihydroxybenzyl alcohol 4,5-disulfate; IV, 2,3-dibromo-4,5-dihydroxybenzyl alcohol 1′,4-disulfate; V, cephamycin A; VI, 1,2,3,5-tetrahydroxybenzene 2,5-disulfate; VII, comantherin sulfate (R = methyl), neocomantherin sulfate (R = propyl); VIII, comaparvin sulfate (Z = H, Y = OH), 6-methoxycomaparvin sulfate (Z = OCH_3, Y = OH), 6-methoxy-5-methylether-comaparvin sulfate (Z = OCH_3, Y = OCH_3); IX, luciferyl sulfate.

I. Algal Aryl Sulfates

Following early work on bromophenols in red algae,[93,94] Hodgkin and colleagues[95] in 1966 reported the isolation from *Polysiphonia lanosa* and characterization of a bromophenyl sulfate to which they assigned the structure in Figure 7,III. Although the same 4,5-disulfate structure has been reported more recently[96] to occur in the marine red alga *Rhodomela subfusa*, the assignment of this structure to the *P. lanosa* compounds has been criticized[97] and rejected in favor of the 1′,4-disulfate structure (Figure 7,IV). This compound which is intriguing in that it incorporates an aryl sulfate and a primary alkyl sulfate in the same molecule, appears to be widely distributed among marine red algae.[97-100]

In addition, it has been suggested[100] that the frequent reports[99,101,102] of unsulfated phenols in algae may be artifacts of the isolation procedure and that these compounds exist in vivo predominantly as sulfate esters. In this context it is relevant to note that whereas the parent phenol 2,3-dibromo 4,5-dihydroxybenzyl alcohol is a potent antibiotic, the disulfate ester is much less effective.[99,103-105] The implications of these observations for microorganisms which produce arylsulfatase enzymes are not clear at the present time. In passing, it should be noted that absence of sulfate is not a general prerequisite for antibiotic activity; thus the antibiotic cephamycin A produced by members of the genus *Streptomyces* is in fact an aryl sulfate ester (Figure 7,V).[106,107]

Brown algae from six orders of Phaeophyceae produce aryl sulfates in the form of the 2-sulfate and 2,5-disulfate of 1,2,3,5-tetrahydroxy-benzene (Figure 7,VI).[108,109] With the application of sensitive spectrophotometric techniques,[110] the discovery of further sources of these and related compounds can be safely anticipated.

J. Naphthoquinone and Naphthopyrone Sulfates

Crinoids, which are marine animals of the phylum *Echinodermata*, are pigmented with linear and angular naphthopyrones. *Comantheria perplexa*[111] produces comantherin- and neocomantherin sulfates (Figure 7,VII), and *Comanthus parvicirrus timorensis*,[112] the angular comaparvin sulfates (Figure 7,VIII). Fusarubin sulfate, a naphthoquinone of uncertain structure,[113] is an aryl sulfate produced by the mold *Fusarium solani*.

Finally, it would be remiss to omit a mention of some of the marine coelenterates that rank among the most beautiful marine creatures in existence. These creatures such as the sea pansy *(Renilla)* and sea feather *(Ptilosarcus)* display a most striking and brilliant luminescence which results from oxidation of the compound luciferin.[114] The storage form of luciferin in these organisms is as the sulfate ester depicted in Figure 7,IX which occurs widely in anthozoans and hydrozoans of the phylum *Cnidaria* and the tentaculata of the phylum *Ctenophora*.[115,116]

Even this incomplete survey illustrates the wide range of aryl sulfate esters which occur naturally. More striking still is the diversity of living organisms which have so far been found to elaborate such compounds, from microorganisms such as fungi and algae, through higher plants, to marine and terrestrial invertebrates and vertebrates. The implication is clear; aryl sulfate esters of diverse structures are returned ubiquitously to soils and natural waters in large quantities. It is therefore not surprising that microorganisms should have evolved to produce arylsulfatase enzymes enabling them to utilize such compounds. Indeed, as will emerge in subsequent sections of this chapter, several naturally occurring aryl sulfates are already known to be hydrolysed by microbial enzymes (for example indoxyl 3-*O*-sulfate, tyrosine-*O*-sulfate, estrone sulfate, and naphthyl sulfate).

III. FUNGAL ARYLSULFATASES

A. Discovery

In 1911, Derrien[117] reported the discovery of a molluscan arylsulfatase which is involved in the conversion of indoxyl sulfate (Figure 3,II) to the Tyrian purple indigo dyes (see Volume I, Chapter 1). This was the first known arylsulfatase enzyme. However, it was not until 1923 that the first fungal arylsulfatase was found[118] in Taka Diastase®, a commercial preparation made by growing the fungus *Aspergillus oryzae* on sterilized mash prepared from crushed cereals, and washing out the soluble mold products after a suitable growth period. The dried extracts were rich in "diastase" (an outdated name for α-amylase) and first came into use at the turn of the century as a treatment for dyspepsia attributed to inadequate digestion of starch. This and other similar preparations have also been used in industry for removal of starch and, because they also contain proteases, peptidases, and lipases, in analytical chemistry for digesting food samples. Indeed, Neuberg has described Takadiastase as "an arsenal of enzymes"[119] and it was in his laboratory that such preparations were shown[118,120-124] to liberate phenols and sulfate from phenyl sulfate, its alkyl-, nitro-, bromo-, and chloro-substituted derivatives, and from aryl sulfate esters found in urine, such as indican (indoxyl 3-*O*-sulfate). As a result, Neuberg and Simon in 1932 suggested the name "phenolsulfatase" for the responsible enzyme.[125]

Since this early work, information has steadily accumulated concerning the substrate specificity and pH optimum of arylsulfatase in Taka Diastase® and other similar commercial preparations of *A. oryzae* (e.g., Clarase, Mylase® P, Polidase® S).[126-139] Although there are a few notable exceptions (see Table 1), these commercial samples exhibit a broad specificity for aryl sulfates and consequently the name "arylsulfatase" was proposed as being more appropriate.[133] This broad specificity, coupled with the enzymes' ability to remain active in the presence of normal and pathological urines,[129,140] lead to their use in the laboratory for the analysis of urinary aryl sulfate esters.[131,135,141]

By 1957, sufficient data had accumulated on arylsulfatases from fungi, bacteria, mollusks, and mammals to enable Dodgson and Spencer[142] to establish a classification of the enzymes based on substrate specificity and the effectiveness of certain inhibitors. Thus Type 1 enzymes exhibit appreciable activity and affinity toward *p*-acetylphenyl sulfate and *p*-nitrophenyl sulfate, but are less active toward 2-hydroxy-5-nitrophenyl sulfate (nitrocatechol sulfate). These enzymes are inhibited by cyanide but are hardly affected by phosphate or sulfate ions. Type 2 enzymes have little activity for *p*-acetylphenyl sulfate and *p*-nitrophenyl sulfate, but considerable activity toward nitrocatechol sulfate. They are unaffected by cyanide, but strongly inhibited by phosphate or sulfate. Although some anomalies have been found and the classification has been criticized for being unnecessarily complex,[143] it has nevertheless proved to be a useful aid in assimilating knowledge of arylsulfatases from a wide variety of sources. The collective evidence (Tables 1 and 2) for Taka Diastase® arylsulfatase enabled Dodgson and Spencer[142] to classify it as a Type 1 enzyme.

Meanwhile, other fungi were coming under scrutiny for potential ability to hydrolyse aryl sulfates. Harada and colleagues were fascinated by the red coloration imparted to newspaper steeped in putrefying urine, and their initial work on the underlying biochemistry lead them to recognize that arylsulfatases of bacteria in the urine were responsible for initiating the process by hydrolysing indoxyl 3-*O*-sulfate present in the urine (see Section IV.A). A fuller account of this phenomenon is presented in the section on bacterial arylsulfatases (Section IV). For the present it is only important that Harada and colleagues undertook to survey the microorganisms which could accomplish this feat, including some fungi. All eight fungal species studied *(Aspergillus*

Table 1
SUBSTRATE SPECIFICITY OF ARYLSULFATASE OF TAKA DIASTASE®

Compound	Hydrolytic behavior	Ref.
Phenyl sulfate	Substrate at optimum pH 6	118,121,128
	Substrate at optimum pH 9	126
	Substrate for "Clarase," "Polidase S," and "Mylase P"	128,132
o-,*m*-,*p*-Cresyl sulfates	Substrates at optimum pH 9 with relative rates *p*> *o*,*m*	120,121,126
	Substrate for "Clarase" enzyme at pH 6	129
Dimethylphenyl sulfate	Substrate at pH 6 for "Clarase" enzyme	129
2-Methyl-5-isopropyl phenyl sulfate	Substrate at pH 6 for "Clarase" enzyme	129
(+) *p-sec* Butylphenyl sulfate	Substrate at pH 6—6.2	123
(−) *p-sec* Butylphenyl sulfate	*Not* a substrate	123
(±) 3-Methylcyclohexylphenyl sulfate	Substrate	124
4-Hydroxybenzaldehyde 4-sulfate	Substrate at optimum pH 5.5—6.1	133,134
2-Carboxyphenyl sulfate	Substrate	127
o-,*m*-,*p*-Nitrophenyl sulfates	Substrates at optimum pH 5.9—6.5	121,130,133,134
	Relative rates	
	m> *p*> *o*	134
	p> *m*> *o*	136
p-Bromophenyl sulfate	Substrate	121
m-,*p*-Chlorophenyl sulfates	Substrates at optimum pH 6.15	121,133,134
o-,*m*-,*p*-Aminophenyl sulfates	Substrates at pH 6 but much slower than nitrophenyl sulfates	136
Aryl sulfates with strongly basic amino group ($pK_a > 4.3$) *ortho* to the sulfate group	*Not* substrates	136
Aryl sulfates with weakly basic amino groups (unprotonated at assay pH 5.8) *ortho* to the sulfate group	Substrates	136
4-Hydroxy-2-nitrophenyl sulfate	Substrate at optimum pH 6	134, 136
2-Hydroxy-5-nitrophenyl sulfate (nitrocatechol sulfate)	Substrate at optimum pH 6	134,136
4-Hydroxy-3-nitrophenyl sulfate	*Not* a substrate	136
Naphthyl 1-sulfate	Substrate at pH 6	122,136
Naphthyl 2-sulfate	Substrate at pH 6	122,136
2-Aminonaphthyl 6-sulfate	Substrate at pH 6	122,136
2-Acetamidonaphthyl 6-sulfate	Substrate at pH 6	122,136
Indoxyl 3-*O*-sulfate	Relatively poor substrate	121,136
	Good substrate for "Clarase" enzyme	129
Quinyl sulfate	Substrate	121
Carvacryl sulfate	Substrate	121
Vanillin sulfate	Substrate	121
Tyrosine-*O*-sulfate	Substrate at optimum pH 6.3	137
Estrone 3-sulfate	Substrate for "Mylase P" enzyme	131,135,138
Dehydroepiandrosterone sulfate	*Not* a substrate	138
Ethyl sulfate	*Not* a substrate	121,136
2-Aminoethyl sulfate	*Not* a substrate	121,136
Pentan 1-yl sulfate	*Not* a substrate	121,136
Mandelic sulfate	*Not* a substrate	121, 136
Phenolphthalein disulfate	Substrate at pH 6.2	139

glaucus, A. oryzae, A. usamii, A. niger, A. awamori, Rhizopus nigricans, Penicillium glaucum, and *Mucor rouxii)* were able to hydrolyse indoxyl 3-*O*-sulfate, although the yeasts *Saccharomyces cerevisiae* and *S. ellipsoideus* did not do so.[144] Further work[145,146] extended the type of aryl sulfate ester which these fungi could hydrolyse to include

Table 2
EFFECT OF CERTAIN ANIONS ON THE ACTIVITY OF ARYLSULFATASE IN CRUDE COMMERCIAL PREPARATIONS OF *A. ORYZAE*

Inhibitor[a]	Substrate	Inhibition (%)	Ref.
Sulfate	*p*-Chlorophenylsulfate	0	134
Sulfate	*p*-Nitrophenylsulfate	0	134
Phosphate	*p*-Chlorophenylsulfate	19	134
Phosphate	*p*-Nitrophenylsulfate	4	134
Phosphate (10 m*M*)	Phenylsulfate	25	132
Cyanide (0.1 m*M*)	Phenylsulfate	85% in 5 min, 100% in 10 min, Irreversible	132
Cyanide (2 m*M*)	*p*-Nitrophenylsulfate	46.5	134
Cyanide	*p*-Nitrophenylsulfate	100	134
Sulfite (10 m*M*)	*p*-Chlorophenylsulfate	68	134
Sulfite (2 m*M*)	*p*-Nitrophenylsulfate	20	134
Sulfite	*p*-Nitrophenylsulfate	75	134

[a] 20 m*M* unless stated otherwise.

phenolphthalein disulfate and *p*-nitrophenyl sulfate. These two esters, especially the latter, proved to be superior substrates to indoxyl 3-sulfate, except for *P. glaucum* and *M. rouxii* which showed low activity to all three compounds. In subsequent studies, Harada and Spencer[147] extended the list of fungi able to hydrolyse aryl sulfates and classified them according to the repression characteristics of arylsulfatase synthesis (see Volume II, Chapter 3). Of 23 active fungi, *A. oryzae* came about half way down the list in terms of ability to liberate *p*-nitrophenol from *p*-nitrophenyl sulfate present in the growth medium. Nevertheless it continued to attract attention as a convenient source of arylsulfatases, and most of the studies that were to follow on the fungal arylsulfatases per se, as opposed to studies on their synthesis, center around this and other *Aspergillus* species.

B. Arylsulfatases of *Aspergillus* Species

1. Multiple Arylsulfatases

A close examination of Tables 1 and 2 shows that by the end of the 1950s, some discrepancies had appeared regarding the properties of *A. oryzae* arylsulfatase in crude preparations, depending on the source of the enzyme and the investigators at whose hands the work was done. For example, although most studies indicated an optimum pH around 6, Fujita and Hosoda[126] reported a much more alkaline value, near pH 9; Boyland and colleagues[136] observed the relative rates of hydrolysis of nitrophenyl sulfate isomers to be in the order *para* > *meta* > *ortho,* whereas Robinson et al.[134] had earlier reported the sequence to be *meta* > *para* > *ortho;* although there is general agreement that cyanide is a strong inhibitor, different sensitivities to this ion have been recorded. With the benefit of hindsight, it is not difficult to see that these clues were probably pointing toward a discovery that was to mark the beginning of a new phase in the study of fungal arylsulfatases; namely that depending on growth conditions and stage of development, fungi produce not one, but a multiplicity of arylsulfatase enzymes. Despite these clues suggesting heterogeneity, attempts to purify the enzyme[148,149] failed to reveal the existence of multiple forms of arylsulfatase activity in fungi until the work of Cherayil and van Kley[150-153] in the early 1960s. Their reports proved to be a turning point from which real progress was to be made in purifying and characterizing the arylsulfatases of *A. oryzae* and of other fungi.

In some initial studies on the optimum conditions for the synthesis of arylsulfatase in *A. oryzae* during growth on moist wheat bran, Cherayil and van Kley[150] noted that the greatest amount of enzyme and highest specific activity were produced when the least amount of water was added to the bran. Readily assimilable sulfur compounds such as sulfate, sulfite, and methionine, markedly repressed arylsulfatase synthesis whereas nitrogen-containing compounds such as ammonium salts and urea, increased production up to 25-fold. Further work[151] showed that the fungus was producing two arylsulfatases distinguishable by different pH optima (5 and 6.3), and that the ratio of the two isoenzymes varies with the amount of water added to the bran, a high moisture content favoring the enzyme with optimum pH at 6.3. The enzymes were also distinguishable by their stability to storage and heating, and in their response to certain inhibitors. The early work[150] had shown that growth in the presence of compounds which inhibit the enzyme activity (such as tungstate and molybdate) lead to enhanced production of arylsulfatase. Further experiments[152] established that this was due to the additional synthesis of a third, minor isoenzyme with an optimum pH of 6.75, higher than for either of the two major isoenzymes. The three arylsulfatases were successfully separated either by electrophoresis or by ion-exchange chromatography on Amberlite® IRC-50. The enzymes were labeled I (optimum pH 6.75), II (optimum pH 5.0), and III (optimum pH 6.3) according to the order of migration on electrophoresis. An examination of the inactivation of the two major isoenzymes II and III by urea and guanidine, and of the effect of urea and dithiothreitol on their molecular weights, led Drnec and van Kley[154,155] to conclude that these isoenzymes are not related by subunit association.

Presslitz[156] extended the work to use a defined liquid culture medium and confirmed that the synthesis of all three arylsulfatases was repressed by sulfate, sulfite, methionine, cysteine, and cystine. Furthermore, when the mold was grown on a medium containing taurine as the source of sulfur, the total activity increased four- to fivefold when tyramine or tyrosine was also present. An initial partial purification by Presslitz[156] of arylsulfatase II and, later, a complete purification of the same isoenzyme by Benkovic's group,[157] has enabled an extensive chemical, physical, and catalytic characterization to be made for this particular enzyme. Further discussion of the properties of arylsulfatase II and its isoenzymes, and of the regulation of their synthesis can be found elsewhere (Sections III.B.2 and Volume II, Chapter 3). However, for the present, attention is confined to the phenomenon of the multiplicity of arylsulfatases in *A. oryzae*.

In view of accumulating evidence that tyramine was an important regulator of the synthesis of arylsulfatases in fungi and bacteria (see Volume II, Chapter 3), Burns and Wynn[158] set out to examine the effect of tyramine on the activity of the enzymes themselves. Using *A. oryzae* they found, like van Kley's group, that three isoenzymes were present and, moreover, that one of these possessed the ability to transfer sulfate from nitrocatechol sulfate to tyramine. The isoenzymes were separated by DEAE-cellulose anion exchange chromatography and designated "I", "II", and "III" in order of elution. The novel sulfotransferase activity which was the subject of subsequent kinetic studies[159] was associated only with fraction "II". However, this is not the same arylsulfatase II which had been discovered by van Kley's group and later studied in Benkovic's laboratory, and quotation marks have been used here to emphasize this point. It appears that Burns and Wynn have unwittingly generated some confusion by introducing a system of notation for *A. oryzae* arylsulfatases which does not coincide with that originally devised by van Kley and later adopted by Benkovic. The latter depends on the order of migration during electrophoresis whereas Burns and Wynn, and more recently Ohtsuki and colleagues,[160] used the order of elution from DEAE anion exchangers. Clearly these do not necessarily correspond.

In order to clarify the situation, some of the data pertaining to the properties of the arylsulfatases, which properly belongs in later sections, has been collected in Table 3. The table is divided according to the two systems of notation and subdivided according to the isoenzymes themselves. Upper case Roman numerals (I, II, III) are retained to denote van Kley's original notation and quotation marks are used to indicate the isoenzymes according to more recent authors.[158,160] It is immediately apparent that the minor component arylsulfatase I which appears when molybdate or tungstate is present in the growth medium is identical with arylsulfatase "III" of Burns and Wynn. This is confirmed by the optimum pH values for these enzymes which are the most alkaline observed by the respective authors. On the basis of optimum pH, molecular weight, and inhibition by phosphate and fluoride ions, there seems also to be little doubt that arylsulfatases II and "I" are identical. By elimination, this leaves arylsulfatase III equivalent to isoenzyme "II". The other data in Table 3 confirm these identities beyond any reasonable doubt. To summarize, the following isoenzymes are equivalent: I and "III", II and "I", III and "II".

It is fortunate that each of the three groups of research workers who have studied the properties of arylsulfatases purified from *A. oryzae* has concentrated on a different isoenzyme. Thus Benkovic's group[157] in the U.S. following van Kley's initial work, purified and characterized the arylsulfatase II; Sakurai et al.[161] working in Japan have recently examined what they called arylsulfatase "III" (isoenzyme I) which exhibits maximum activity at alkaline pH; finally, Burns and Wynn[158] in the U.K. found the sulfotransferase activity associated with arylsulfatase "II", that is isoenzyme III. The last authors have also devised an assay procedure for determination of the activities of each of these isoenzymes[162] which is based on their different activities towards phenolphthalein disulfate and *p*-nitrophenylsulfate in the presence and absence of tyramine. The purification and general properties of each isoenzyme is now described in the order given, which is the chronological order in which they were studied. Because neither system of notation offers any particular advantage over the other, that due to Cherayil and van Kley and later adopted by Benkovic's group is used in this account simply because it was the first.

The physiological purpose of a multiplicity of arylsulfatases in *A. oryzae* is a most intriguing question. However, further examples of arylsulfatase multiplicity will emerge for fungal, bacterial, and algal sources, and because an analysis of this problem is facilitated by a knowledge of the properties of the individual enzymes, this matter is deferred until these properties have been described in more detail.

2. *Arylsulfatase II of A. oryzae*

a. *Purification*

Initial studies in van Kley's laboratory which first showed the existence of arylsulfatase isoenzymes in *A. oryzae* also established an outline procedure for the purification of arylsulfatase II. For this purpose, the mold was grown on wheat bran moistened with a solution containing urea and diammonium hydrogen phosphate, conditions which produce high levels of arylsulfatase II.[150,153] This enzyme was then partly purified (200-fold) from extracts of the bran mold by a combination of acetone precipitation, IRC-50 chromatography (a weak cation-exchange resin containing −COOH groups), and zone electrophoresis.[153] Drnec[154] later improved upon this basic purification method by inserting some extra steps. The final procedure which consisted of acetone precipitation, DEAE-Sephadex®, SE-Sephadex® and IRC-50 column chromatography steps, polyacrylamide gel electrophoresis, and finally Sephadex® G-100 gel filtration, enabled a complete purification of arylsulfatase II. The final preparation was homogeneous on the basis of sedimentation velocity ultracentrifugation and gave a single enzymatically active protein band on polyacrylamide gel electrophoresis. These

Table 3
NOTATIONS FOR ARYLSULFATASE ISOZYMES IN *A. ORYZAE*

Property	Notation[a] of Cherayil and van Kley			Notation[b] of Burns and Wynn		
	I	II	III	"I"	"II"	"III"
Relative amount synthesized	Minor, induced by MoO_4^{2-} or WO_4^{2-}	Major component	Major component	Major component	Major component	Minor, induced by MoO_4^{2-} or WO_4^{2-}
pH optimum	6.75 (152)[c]	5.0 (152) 4.8 (151)	6.3 (151,152)	5.0 (160)	6.7 (160)	8.0 (160)
Molecular weight	—	95,000 (164)	>90,000 (155)	105,000 (158)	140,000 (158)	57,000 (158)
Inhibition by phosphate	46%[d] (153)	60%[d] (153) 44% (157)	12.5%[d] (153)	55.5% (160)	12.5% (160)	—
Inhibition by fluoride	76%[d] (153)	90%[d] (153) 69% (157)	48%[d] (153)	82.5% (160)	9% (160)	—
Stability at various pH	—	Stable at pH 5—8.5 (151)	Less stable at pH 5—8.5 (151)	Stable at pH 5—8.5 (160)	Stable at pH 5—8.5 (160)	Stable at pH 8—10, unstable at pH 4.5—8 (160)
Stability to heating	—	II more stable than III		"I" more stable than "II"		—

[a] Based on order of migration during electrophoresis.
[b] Based on order of elution from DEAE-cellulose.
[c] Numbers in parentheses indicate reference numbers.
[d] Marked values refer to 5 m*M* *p*-nitrophenyl sulfate and 5 m*M* inhibitor. Unmarked values refer to 1 m*M* concentrations of each.

methods remained unpublished (except in abstract form) until Benkovic's group[157] developed their interest in the enzyme. These workers employed basically the same method except that only one of the modifications that Drnec had introduced was used, namely the Sephadex® gel-filtration step. Thus a sequence of acetone precipitation, BioRex® 70 chromatography (a weak cation exchanger equivalent to IRC-50), Sephadex® G-100 gel filtration, and preparative gel electrophoresis achieved a 760-fold purification of arylsulfatase II from the original bran mold, with a recovery of 7.8% of the activity. The final preparation was homogeneous on analytical polyacrylamide gel electrophoresis, sucrose density gradient centrifugation, and ultracentrifugation, with coincidence of protein and enzyme activity in all cases.[157]

Both groups reported that arylsulfatase II was stable to storage around pH 5.0 and that the enzyme was suitable for use in physical and kinetic studies without further treatment. In fact, this was the most stable of the three *A. oryzae* arylsulfatases, which might account for it having attracted the most attention.

b. Properties

The literature contains several different values for the molecular weight of arylsulfatase II of *A. oryzae* (see Table 4). Generally speaking, gel-filtration methods have given higher values (in the 84,000 to 105,000 range) than the earlier sedimentation velocity or sedimentation equilibrium experiments (65,000 to 75,000). Drnec[154] investigated the possibility of a subunit structure for arylsulfatase II with a view to establishing a possible relationship with its isoenzymes based on subunit hybridization. Although the enzyme was reversibly inactivated by urea or guanidine hydrochloride, the effect was probably not due to dissociation of noncovalently bound subunits since elution volumes of the protein on Sephadex® G-150 were the same in both the absence and the presence of urea. The presence of 8 *M* urea also did not affect the electrophoretic mobility of the enzyme, which always migrated as a single, symmetrical protein band. Similar results were obtained when the enzyme was subjected to gel filtration chromatography in the absence and in the combined presence of dithiothreitol and guanidine hydrochloride. Evidently no change in protein size had occurred, and Drnec and van Kley[155] were thus led to conclude that arylsulfatase II was a single polypeptide chain.

There it seemed the matter would rest until Benkovic and Fedor[163] developed an active site titrant for arylsulfatase II. This compound, *o*-nitrophenyloxalate, inactivates the arylsulfatase II, and activity measurements made during the titration of the protein with inhibitor indicated that the completely inactivated protein would contain 2.1 oxalate ester residues per molecule of enzyme. The collective results implied, but did not require, the presence of two active sites per molecule of enzyme. These experiments renewed interest in the possibility of a subunit structure for the enzyme and lead Benkovic's group to conduct a more detailed examination of the molecular weight in the presence of various dissociating agents.[164] Enzyme exhaustively dialyzed against buffer, pH 4.0, appeared to be homogeneous on polyacrylamide gel electrophoresis unless urea or sodium dodecyl sulfate was also present during dialysis, in which case two bands appeared corresponding to molecular weights of 40,000 and 86,000. The implied dimeric structure for the native enzyme was confirmed by sedimentation equilibrium experiments which, in the absence of urea at pH 4.83, gave a native molecular weight of 94,900, but a value of 53,200 in the presence of urea at pH 4.04. However, dissociation of subunits appeared to depend on pH, because polyacrylamide gel electrophoresis of samples pretreated with urea or sodium dodecyl sulfate at a higher pH (7.6) revealed a single band. In addition, sedimentation equilibrium centrifugation at pH 7.6 in the presence of urea produced nonlinear data which was indicative of only partial dissociation into subunits. It thus appears that the enzyme consists of two iden-

Table 4
MOLECULAR WEIGHTS OF MICROBIAL ARYLSULFATASES

Organism	Isoenzyme	Molecular weight	Method	Ref.
Aspergillus oryzae	I ("III")[a]	57,000	Gel filtration	158
	II ("I")[a]	84,000	Gel filtration	154
		75,300	Sedimentation velocity	154
		90,000—100,000	Gel filtration	155
		65,000	Sedimentation equilibrium	157
		105,000	Gel filtration	158
		94,900	Sedimentation equilibrium pH 4.83	164
		91,300	Sedimentation equilibrium pH 7.61	164
		53,200	Sedimentation equilibrium pH 4.04 in the presence of 8 *M* urea	164
		86,000[b]	SDS-gel electrophoresis	164
		40,000[b]		164
	III ("II")[a]	88,000	Gel filtration	154
		>90,000	Gel filtration	155
		140,000	Gel filtration	158
A. sojae	Alkaline enzyme	60,000	Gel filtration	161
A. awamori	Alkaline enzyme	60,000	Gel filtration	161
A. nidulans	I	44,000	Sedimentation velocity	168
	II	44,000	Sedimentation velocity	168
Cephalosporium acremonium	—	50,000	Gel filtration	192
Aerobacter aerogenes	—	40,700	Sedimentation equilibrium	221
Klebsiella aerogenes	—	47,000	SDS-gel electrophoresis	224
		45,000	Gel filtration	224
Serratia marcescens	—	46,000	SDS-gel electrophoresis	234
		49,000	Gel filtration	234
Pseudomonas aeruginosa	α	60,000	SDS-gel electrophoresis	254
	β	60,000	SDS-gel electrophoresis	254
Pseudomonas C12B	—	50,783	SDS-gel electrophoresis	261
		52,318	Sedimentation velocity	261
Chlamydomonas reinhardti	—	150,000	Gel filtration	288
		152,000	Sedimentation velocity	288

[a] See Section III.B.1 for explanation of notation.
[b] Two bands observed on gel zymograms.

tical subunits that separate in the presence of urea or sodium dodecyl sulfate at pH 4.0 to 5.0, but not at pH 7.6. Sodium dodecyl sulfate was a simple competitive inhibitor of the enzymic hydrolysis at pH 7.6 (in contrast to the nonlinear inhibition observed at pH 4.04) and in view of the low isoelectric point of his enzyme[154,157,164] (around pH 4.3) it was considered likely[164] that the high negative charge on the protein at pH 7.6 might prevent binding of dodecyl sulfate at sites other than the active site. The use of an inappropriate pH might account for the failure of previous investigators to separate the subunits. Other examples are known of the importance of pH effects in the subunit dissociation of sulfatase enzymes. The effect of sodium dodecyl sulfate on ox liver arylsulfatase A, for example, depends on the net charge on the molecule (pI = 3.6). The enzyme is a monomer at pH 7.5 with a molecular weight of 107,000 and a tetramer at pH 5.0 (411,000). At the appropriate pH, sodium dodecyl sulfate dissociates the tetramer, but not the monomer, into smaller identical subunits (molecular weight 24,000).[165]

Before leaving this section on molecular weights it should be noted that the comparison which Sampson et al.[164] make between their estimates of the molecular weight of arylsulfatase II and that reported by Rasburn and Wynn[166] for an isoenzyme which they called "classical" arylsulfatase should not be taken as evidence for identity for two reasons. First, the latter preparation had a pH optimum at 6.3 and therefore is most likely to be arylsulfatase III in van Kley's notation (see Table 3). Second, the molecular weight obtained by Rasburn and Wynn was quoted as 100,000 on the basis of its elution from a Sephadex® G-100 gel filtration column. Clearly the enzyme must be very near the exclusion limit of Sephadex® G-100 and estimates of its molecular weight obtained in this way must be treated with caution.

Benkovic et al.[157] reported the amino acid composition of the purified enzyme and calculated the number of residues of individual amino acids in the enzyme molecule assuming a molecular weight of 65,000. In view of more recent estimates of molecular weight, these figures are probably in need of revision. However, the salient features of the analysis remain unchanged. The enzyme is rich in glycine, aspartate, and glutamate, and in this respect it again resembles arylsulfatase A of ox liver.[167] Whether or not this similarity is reflected in common sequences of amino acids in these and indeed other arylsulfatases remains to be investigated, but it is an interesting possibility. The presence of at least four cysteine residues[157] (and probably nearer six when calculated using a revised molecular weight) may be related to the decreased enzyme activity observed in the presence of dithiothreitol.[154] This compound probably acts by reducing cystine S–S bonds. If indeed such S–S bonds are present, they most likely occur as intra- rather than intersubunit links because Sampson et al.[164] achieved subunit separation without the need for a reducing agent. It is therefore perhaps not surprising that treatment with dithiothreitol decreases the activity of the enzyme but does not separate subunits.[154]

Arylsulfatase II of *A. oryzae* like all other known arylsulfatases (see Volume II, Chapter 2) catalyzes the hydrolytic fission of the O–S bond of aryl sulfate esters to liberate the parent phenol and sulfate.[164] As might have been expected from the broad substrate specificity of crude preparations such as Taka Diastase® (Table 1), the range of aryl sulfate esters which serve as substrates is wide. Benkovic et al.[157] reported the kinetic parameters for a Michaelis-Menten mechanism for the hydrolysis of a range of aryl sulfate esters, mono- or disubstituted in the benzene ring with groups known to influence electron withdrawal into the aryl nucleus. V_{max} was essentially independent of the electron-withdrawing power of the ring whereas K_m was generally lower when electron-withdrawing substituents were present. These results were interpreted as strongly suggesting, but not mandating, that a sulfuryl-enzyme intermediate was formed, the rate of hydrolysis of which was the overall rate-limiting step.[164] However, further evidence based on stopped-flow kinetics, partitioning experiments in the presence of amine nucleophiles, and ^{18}O-exchange studies was to show that this was not the case. Further discussion of the mechanism of action of this and other arylsulfatases and a comparison with established mechanisms of nonenzymic hydrolysis can be found in Volume II, Chapter 2. Burns and Wynn showed that arylsulfatase II (referred to by these authors as arylsulfatase "I") was distinguished from the others by its ability to hydrolyse both tyrosine-*O*-sulfate[158] and phenolphthalein disulfate.[162] The latter compound was therefore used to assay specifically for arylsulfatase II in the presence of the other isoenzymes.

At this point it is useful to relate the substrate specificity of arylsulfatase II to some of the naturally occurring aryl sulfates described at the beginning of this chapter. Dodgson et al.[137] demonstrated that Taka Diastase® could hydrolyse tyrosine-*O*-sulfate and as just mentioned Burns and Wynn[162] showed that of the three arylsulfatases present in *A. oryzae,* only isoenzyme II was active in this respect with K_m = 25 m*M* at pH 5.9. At about the same time, Sampson et al.[164] failed to detect any inhibition of arylsulfatase II by tyrosine-*O*-sulfate at pH 4.0 or 7.6, and therefore did not proceed to test it as a substrate. However, a peptide containing tyrosine-*O*-sulfate (*N*-carbobenzoxy-L-phenylalanyl-L-tyrosine-*O*-sulfate methyl ester) was a competitive inhibitor of *p*-nitrophenyl sulfate hydrolysis, and, moreover, itself served as a substrate (K_m = 1.37 m*M* at pH 4.0). The desulfated dipeptide was neither an inhibitor nor a substrate. Apparently the active site does not readily accommodate the zwitterionic tyrosine-*O*-sulfate, but when α-amino and α-carboxyl groups are modified with phenylalanyl and methyl moieties, respectively, binding can then occur. In keeping with this observation, indoxyl 3-*O*-sulfate is not inhibitory at pH 4.0, but gives mixed inhibition at pH 7.6. Assuming an approximate pK_a of 4.9 for the indole nitrogen atom, indoxyl 3-*O*-sulfate was reckoned to be protonated at pH 4.0 and accordingly possesses a low affinity for the enzyme. The early experiments of Boyland et al.[136] with Taka Diastase® also showed that *o*-aminoaryl sulfates bearing amino groups which were protonated under the assay conditions were not hydrolysed. Aliphatic and monosaccharide sulfates and phosphates, and the nonaromatic steroid ester, androstenolone 3β-sulfate failed to bind to the purified enzyme.[164] Estrone 3-sulfate, on the other hand, acts as an inhibitor of *p*-nitrophenyl sulfate hydrolysis, and as a substrate itself with $K_i \approx K_m \approx 0.07$ m*M* (pH 4.0) a value considerably below that for *p*-nitrophenyl sulfate.[164] Thus the enzyme apparently requires an aromatic sulfate ester whose binding is enhanced by incorporating the aromatic moiety in a hydrophobic backbone, but is decreased by the presence of positively charged substituents.

One of the characteristic features which serves to distinguish arylsulfatase II from the other *A. oryzae* isoenzymes (Tables 3 and 5) is its acid pH optimum[151,152,157,160] around pH 5.0. Sampson et al.[164] described the pH-dependence of the overall reaction in terms of the effect of pH on K_m and V_{max}. High V_{max} was dependent on the presence of the base form of a group with pK_a = 4.7 and the acid form of a group with pK_a = 6.3. On the other hand, K_m depends only on a group with pK_a = 6.1 with lower values being found on the acid side. It was considered likely that below pH = 4.0 the decrease in V_{max} may correspond to dissociation into inactive or partially active subunits. However there is no evidence to support subunit formation on the basic side.

The groups affecting V_{max} and K_m with pK_a's of 6.3 and 6.1, respectively, may be one and the same moiety. On the basis of pH effects on Michaelis-Menten parameters, Presslitz[156] has suggested that histidine and an α-amino group may be involved in the active center of the enzyme. Treatment with group-specific reagents provided additional evidence for an α-amino group, and also suggested that a tyrosine residue may feature in the hydrolysis. In contrast, the enzyme does not depend for its activity on the presence of sulfhydryl or disulfide units and, indeed, the latter has already been

Table 5
FUNGAL ARYLSULFATASE ISOENZYMES AND THEIR OPTIMUM pH VALUES

Organism	Isoenzyme	Optimum pH[a]	Ref.
Aspergillus oryzae	I ("III")[b]	6.75, 8.0	152,153,170
	II ("I")[b]	4.8, 5.0	151,152,157,160
	III ("II")[b]	6.3, 6.7	151,152,158,160
A. sojae	Alkaline arylsulfatase	8.5	161
A. awamori	Alkaline arylsulfatase	8.5	161
A. nidulans	I	6.5—8.7	168
	II	6.5—8.7	168
Cephalosporium acremonium	—	5.9 (acetate buffer) 7.0 (tris buffer)	192
Fomes geotropus	—	7.5	193
Neurospora crassa	Conidial "patent" enzyme	7.6—7.8	172
	Conidial "cryptic" enzyme	7.6—7.8	172

[a] Using *p*-nitrophenyl sulfate.
[b] See Section III.B.1 for explanation of notation.

excluded from any involvement in subunit association.[156] For further discussion of the mechanism of action of arylsulfatases, the reader is referred to Volume II, Chapter 2.

Arylsulfatase II activity was unaffected by treatment with 0.01 *M* ethylenediamine tetraacetate.[157] However, while there is general agreement that metal ions are not required as activators of the enzyme,[157,160] there are conflicting reports of the ability of metal ions to inhibit (Table 6). Thus Benkovic et al.[157] discovered that Pb^{2+} and Al^{3+} ions produced inhibition (50% at 1 m*M*) while Zn^{2+}, Ca^{2+}, and Cu^{2+} ions were less effective (30% inhibition) at the same concentration, and the monovalent Na^{+}, K^{+}, and NH_4^{+} ions were not inhibitory at all. Ohtsuki and colleagues[160] more recently observed no inhibition by Na^{+}, K^{+}, Zn^{2+}, Ca^{2+}, Cu^{2+}, or Al^{3+} ions and only Hg^{2+} ions were inhibitory (50% at 1 m*M*). The reason for these discrepancies is unknown.

Certain inorganic anions are also inhibitors and here there is a closer consensus as to their effects. Both Benkovic et al.[157] and Ohtsuki et al.[160] recorded about 50% inhibition in the presence of 1 m*M* phosphate and 70 to 80% with 1 m*M* fluoride ions (Table 7). Although early work suggested no inhibition by sulfate, this was later revised when it was found to be a weak competitive inhibitor (K_m = 12.5 m*M*) at pH = 5.0, but not inhibitory at higher pH. This behavior is discussed in relation to mechanism of action and to the use of sulfate inhibition as a criterion for deciding a Type 1 or Type 2 classification in Volume II, Chapter 2. Cyanide[157] and chloride[157] were without effect on arylsulfatase II, but sulfite,[157] borate,[160] tungstate,[160] and molybdate[160] were all strongly inhibitory. On the basis of the effects of some of the inhibitors just described and the relative rates for *p*-nitrophenyl sulfate and nitrocatechol sulfate hydrolysis, Presslitz[156] and Benkovic et al.[157] independently showed that arylsulfatase II of *A. oryzae* is neither a Type 1 nor a Type 2 enzyme.

3. Arylsulfatase III of A. oryzae

As emphasized in Section III.B.1, this isoenzyme, first discovered in crude extracts of *A. oryzae* by Cherayil and van Kley,[150,153] is almost certainly the same isoenzyme referred to as arylsulfatase "II" described more recently by Burns and Wynn[158] and Ohtsuki et al.[160]

a. Purification

Cherayil[153] achieved a 100-fold purification of arylsulfatase III which was separable from isoenzymes I and II by IRC-50 ion-exchange chromatography or zone electropho-

Table 6
EFFECT OF CATIONS ON MICROBIAL ARYLSULFATASES

Source of enzyme	Li^+	Na^+	K^+	NH_4^+	Ag^+	Mg^{2+}	Ca^{2+}	Sr^{2+}	Zn^{2+}	Cu^{2+}	Pb^{2+}	Hg^{2+}	Sn^{2+}	Cd^{2+}	Co^{2+}	Fe^{2+}	Mn^{2+}	Al^{3+}	Fe^{3+}	Tyn[a]	Ref.
Aspergillus oryzae I	·	·	·	·	·	·	·	·	·	·	·	·	·	·	·	·	·	·	·	+	158
A. oryzae II	·	0	0	0	·	0	—	·	—	—	—	—	0	0	—	·	·	—	·	0	157,158,160
A. oryzae III	·	0	0	·	·	0	0	·	0	0	·	—	—	0	0	·	·	0	—	‡	157,158,160
A. sojae	0	0	0	·	=	0	0	0	=	0	·	=	·	—	0	—	0	—	·	·	161
A. awamori	0	0	0	·	=	0	0	0	=	0	·	=	·	—	0	—	0	—	·	·	161
Cephalosporium acremonium	·	·	0	0	·	0	0	·	0	0	·	·	·	·	·	0	·	·	·	·	192
Aerobacter aerogenes	0	0	0	‡	0	+	+	·	—	=	—	·	·	0	0	0	+	·	0	0	218,220,221
Klebsiella aerogenes	·	0	0	+	·	0	0	·	·	·	·	·	·	·	·	·	0	·	·	=	224
Salmonella typhimurium	·	·	·	·	·	·	·	·	·	·	·	·	·	·	·	·	·	·	·	=	
Serratia marcescens	·	·	·	·	·	·	·	·	·	·	·	·	·	·	·	·	·	·	·	=	234
Proteus vulgaris	·	·	·	·	·	0	·	·	—	·	·	·	·	·	·	·	0	·	·	·	235
Alcaligenes metalcaligenes	·	0	·	·	·	·	·	·	·	·	·	·	·	·	·	·	·	·	·	·	244
Pseudomonas C12B	·	·	·	·	·	·	·	·	·	·	·	·	·	·	·	·	·	·	·	+	260,262

Note: ‡ strong activation; +, weak to moderate activation; 0, no effect; —, weak to moderate inhibition; =, strong inhibition; ·, not determined.

[a] Tyramine.

Table 7
EFFECT OF ANIONS ON FUNGAL AND ALGAL ARYLSULFATASES

Source of enzyme	PO_4^{3-}	SO_4^{2-}	CN^-	Cl^-	F^-	SO_3^{2-}	$S_2O_3^{2-}$	Ac[a]	AsO_3^-	AsO_2^-	N_3^-	BO_3^{3-}	WO_4^{2-}	MoO_4^{2-}	EDTA[b]	Ref.
Aspergillus oryzae I	–	O	–	·	–	·	·	·	·	·	·	·	–	–	·	150—153
A. oryzae II	–	O	O	O	=	=	·	·	·	·	–	=	=	=	O	157,160
A. oryzae III	O	·	O	O	·	·	·	·	·	·	–	=	=	=	·	153,160
A. sojae	–	O	O	·	–	·	·	·	O	O	O	=	O	O	O	161
A. awamori	–	O	O	·	–	·	·	·	O	O	O	=	O	O	O	161
A. nidulans I	–	–	=	·	–	=	–	·	–	–	·	·	·	·	·	168
A. nidulans II	O	–	=	·	O	=	O	·	O	O	·	·	·	·	·	168
Cephalosporium acremonium	O	O	=	O	–	·	·	·	·	·	O	·	·	·	·	192
Chlamydomonas reinhardti	O	O	=	·	·	O	·	·	·	·	·	=	·	·	O	288

Note: O, no effect; –, weak to moderate inhibition; =, strong inhibition; ·, not determined.

[a] Acetate.
[b] Ethylenediamine tetra-acetate.

resis. Drnec[154] later reported some properties of a partly purified arylsulfatase III which was presumably prepared in the same way. No further progress was made until 1973 when Wynn and colleagues developed an interest in the *A. oryzae* arylsulfatase isoenzymes. Rasburn and Wynn[166] described the purification in good yield of an isoenzyme which they referred to as "classical" arylsulfatase of *A. oryzae*, and the chemical and physical properties of which were distinct from those of arylsulfatase II of Benkovic et al. Indeed, the pH optimum at 6.3 suggests that this is probably arylsulfatase III in van Kley's notation (Table 3). Rasburn and Wynn[166] used Taka Diastase® as starting material and from it they purified arylsulfatase III by a combination of ammonium sulfate fractionation (70 to 90% saturation at pH 6.3), Sephadex® G-100 and DEAE-cellulose chromatography. Except for some minor contaminants the final preparation was homogeneous when examined by polyacrylamide gel electrophoresis and end-group analysis. The overall purification was 166-fold and 21% of the total activity was recovered.

By 1975, evidence had steadily accumulated of the involvement of tyramine in the regulation of bacterial and fungal arylsulfatases (Volume II, Chapter 3). These observations prompted Burns and Wynn to investigate the effect of tyramine on the activity of the enzymes themselves and they chose *A. oryzae* as an organism suitable for study. Preliminary experiments[158] indicated that tyramine greatly enhanced the arylsulfatase activity in crude extracts and because by this time the occurrence of multiple arylsulfatases in microorganisms was well established, Burns and Wynn resolved to separate the isoenzymes to discover the specificity of the tyramine effect. Extracts of acetone-dried powders of *A. oryzae* were applied to a DEAE-cellulose column and eluted with a salt-gradient. Three distinct peaks of activity were observed (labeled fractions "I", "II", and "III" in order of elution) although there was some overlap of the last two. The enhancement of activity by tyramine was found for fraction "II" and to a much lesser extent for fraction "III". Each fraction was further purified by Sephadex® G-100 chromatography which effected a complete separation of fractions "II" and "III". The report contains no description of tests for homogeneity but, in view of the previous work by Rasburn and Wynn[166] using very similar procedures, it is reasonable to suppose that high levels of purification had been achieved. The nonalignment of the notation of Burns and Wynn with that introduced earlier by van Kley and colleagues has already been dealt with in Section III.B.1 from which the reader will recall that it is the sulfotransferase-associated fraction "II" of Burns and Wynn that is almost certainly identical with arylsulfatase III in the earlier notation.

More recently Ohtsuki et al.[160] have used a similar procedure to separate *A. oryzae* isoenzymes. Following an initial ethanol precipitation step, these workers achieved a separation of the arylsulfatases into two fractions using DEAE-cellulose. Like Burns and Wynn, these they labeled fractions "I" and "II" (i.e., II and III, respectively, in van Kley's notation, Table 3). The third isoenzyme did not appear in quantity unless molybdate was also present (see the following section). Fractions "I" and "II" were further purified by ECTEOLA-cellulose chromatography to produce purifications of 40- and 32-fold in yields of 36 and 28%, respectively. Clearly, the high yields were obtained by these workers at the expense of lower purifications. Arylsulfatase III is somewhat less stable than isoenzyme II with respect to storage and to elevated temperatures. It is also particularly unstable at pH less than 5.5 and is best stored in the cold at pH above this value.[160]

b. Properties

Arylsulfatase III was one of the two major isoenzymes first discovered by Cherayil and van Kley[150] and is distinguishable from the other major isoenzyme by its optimum pH in the range 6.3 to 6.7. It has since received considerable attention particularly recently in the laboratory of Wynn, who refers to it as arylsulfatase "II". Although Drnec[154] observed a molecular weight of 88,000 to 90,000 from gel filtration experiments using Sephadex® G-100, Burns and Wynn[158] recorded a considerably higher value of 140,000 using Sephadex® G-200. It will be recalled that early estimates of the molecular weight of isoenzyme II have since been found to be in need of upward revision (Table 4) and this also appears to be the case for arylsulfatase III. The arylsulfatase described by Rasburn and Wynn[166] with optimum pH 6.3, which is most probably isoenzyme III, was assigned a molecular weight of 100,000, but this value must be considered to be a lower limit because it was estimated from elution volumes on Sephadex® G-100. Inhibition experiments with guanidine HCl indicated that arylsulfatases II and III were not related by differences in noncovalent binding among subunits.[154] This conclusion was based partly on failure to observe dissociation of isoenzyme II into subunits. However, recent success of Benkovic's group in establishing the dimeric nature of this sulfatase suggests that a reappraisal of possible subunit relationships among *A. oryzae* arylsulfatases, would be appropriate, particularly in the light of the possible close relationship between isoenzymes in *A. nidulans* (see Section III.B.5.).

Ohtsuki et al.[160] have compared some general properties of arylsulfatases II and III, in particular their susceptibility to inhibition by metal ions and inorganic anions. Arylsulfatase III, like isoenzyme II, was generally insensitive to a range of metal ions. It was less sensitive than arylsulfatase II to inhibition by Hg^{2+}, but more sensitive to Sn^{2+}, Cd^{2+}, and Fe^{3+} although the inhibition was still slight (Table 6). In contrast to arylsulfatase II which is strongly inhibited by fluoride and phosphate, these ions have little effect on arylsulfatase III. On the other hand, both enzymes are strongly inhibited by molybdate, tungstate, and borate ions (Table 7).

By far the most fascinating feature of arylsulfatase III is its ability to catalyze the transfer of sulfate from arylsulfates to acceptors other than water. Mention has already been made of this phenomenon (Section III.B.1) that was discovered by Burns and Wynn[158] who found that the presence of 20 m*M* tyramine caused an 18-fold increase in the rate of nitrocatechol release from nitrocatechol sulfate. Other phenols could also serve in a similar capacity, but there seemed to be little correlation between structure and ability to enhance nitrocatechol release. Tyramine was in fact among the least effective phenols, with *p*-hydroxyphenyl propionic acid and 3-methoxy-4-hydroxybenzaldehyde being the most active. Little further information could be drawn from the data except that the specificity of the effect was quite broad.

The possibility was then examined that enhancement of arylsulfatase activity arose from the ability of added phenols to act as acceptors of sulfate in a transfer reaction. Incubation mixtures containing arylsulfatase III, nitrocatechol sulfate, and tyramine were analyzed for release of both sulfate and nitrocatechol. While the phenol was released at a steady rate, there was virtually no release of inorganic sulfate into the assay medium. When arylsulfatase II was examined in the same way, nitrocatechol and inorganic sulfate were indeed released together in stoichiometric amounts. Incubation mixtures containing arylsulfatase III, nitrocatechol sulfate, and one of several added phenols were examined by thin-layer chromatography and in each case the corresponding sulfate ester of the added phenol was detected, showing that transfer had occurred. When $^{35}SO_4^{2-}$ was also present, there was no incorporation of radiolabel into any sulfate esters, and the reaction therefore does not proceed via the formation of free inorganic sulfate. PAPS, the usual donor in biological sulfation reactions, was unable to replace nitrocatechol sulfate, neither did it have any effect on the hydrolysis of *p*-nitrophenyl sulfate, transfer of sulfate from *p*-nitrophenyl sulfate to tyramine, or transfer from *p*-acetylphenylsulfate to phenol. These properties distinguish the enzyme quite clearly from mammalian sulfotransferase enzymes which depend absolutely on

PAPS as sulfate donor. It was suggested that in vivo the enzyme probably functions as a sulfotransferase, although in the absence of externally added acceptors for the sulfate group, it will function in vitro as a sulfatase.[158]

The sulfotransferase activity later formed the subject of a kinetic study.[159] Using *p*-nitrophenyl sulfate as sulfate donor and tyramine as acceptor, linear double reciprocal plots were obtained and the overall pattern was consistent with a Ping-Pong® mechanism. Values of K_m for *p*-nitrophenyl sulfate and tyramine at saturating concentrations of the cosubstrate were 4.3 and 2.0 m*M*, respectively. The kinetics provided strong evidence for a covalent enzyme-sulfate intermediate, and the arylsulfatase activity was interpreted as a transfer of sulfate from enzyme-sulfate complex to water rather than a phenolic acceptor. This and other mechanistic aspects are discussed further in Volume II, Chapter 2.

Kinetics of the sulfatase activity of isoenzyme III were also examined.[159] *p*-Nitrophenyl sulfate was hydrolysed in accord with Michaelis-Menten kinetics. However, nitrocatechol sulfate produced nonlinear double reciprocal plots particularly at low concentrations. Examination of incubation mixtures by thin-layer chromatography showed the presence of three aromatic compounds, namely nitrocatechol sulfate, nitrocatechol, and nitrocatechol disulfate. Evidently the reaction is not purely hydrolytic when nitrocatechol sulfate alone is present because the free phenolic hydroxyl group of nitrocatechol sulfate can itself serve as acceptor. This behavior leads to a rate equation containing squared terms in substrate concentration and this was shown to account satisfactorily for the observed nonlinearity in the kinetics.[159]

The possible roles of a sulfotransferase enzyme of this type are considered in Volume II, Chapter 4. Meanwhile it remains to be seen how widespread is this phenomenon. In any event, as Burns and Wynn pointed out,[158] it is clearly important that investigations of the ability of tyramine and other phenols to affect the synthesis of arylsulfatases in microorganisms should take account of the potential effects on enzyme activity of such phenols which may be present in cell extracts and thereby transferred to assay mixtures.

4. *Arylsulfatase I of A. oryzae and the Related Alkaline Arylsulfatases of A. sojae SH 10-6 and A. awamori R-0827*

Cherayil[153] noticed that when molybdate or tungstate was present in the growth medium used to moisten wheat bran on which *A. oryzae* was growing, the synthesis of arylsulfatase was enhanced. The effect was traced to the production of a third isoenzyme with optimum pH 6.75 in addition to the major isoenzymes II and III. This interesting enzyme is in fact the least well characterized of the *A. oryzae* arylsulfatases and it has been neglected until quite recently when Ohtsuki et al.[160] have taken an interest. These workers found that the pH optimum of arylsulfatase activity in cell extracts of *A. oryzae* shifted from pH 6 to 8 when molybdate ions were present in the medium used to moisten the wheat bran supporting growth. Moreover, the pH activity curve of these extracts was identical in shape to that observed for extracts from fungi grown in the absence of molybdate but assayed in its presence. It therefore appears that molybdate induced the synthesis of an isoenzyme with pH optimum at 8.0 which, in contrast to the other isoenzymes, is insensitive to inhibition by molybdate. The reason why this isoenzyme exhibited a somewhat higher pH optimum than that observed by van Kley's group is unknown.

Several other *Aspergillus* fungi were examined[160] and found to synthesize arylsulfatases which, when assayed together in crude extracts, produced pH curves very similar to those observed for *A. oryzae* including a shoulder at around pH 8.0. When extracts were examined for arylsulfatase activity at pH 8.5, all showed considerable activity with particular strains of *A. sojae* and *A. awamori* (a black *Aspergillus* used in the

Japanese fermentation industry) producing highest activity.[161] These organisms were consequently used as sources for the purification and characterization of the alkaline arylsulfatases which are described in the following section.

When *A. sojae* SH10-6 and *A. awamori* R-0827 were cultured on defatted soybean meal or wheat bran, alkaline arylsulfatase activity (assayed at pH 8.5) reached maximum levels after 1 to 3 days depending on the temperature of the culture (25, 30, or 35°C).[161] Activity decreased thereafter and most rapidly at the highest temperature. The results suggested that alkaline arylsulfatase is produced in association with fungal growth, but that the activity is unstable above 30°C. It is interesting that no addition of molybdate or tungstate appears to have been necessary for the production of this enzyme as was the case in earlier studies.

a. *Purification*

Crude enzyme extracts were prepared from *A. sojae* and *A. awamori* grown separately on soybean meal medium at 30°C. In both cases the following purification procedure was used.[161] After ammonium sulfate fractionation at pH 9.0, extracts were subjected to DEAE-cellulose column chromatography. In each case the column was eluted with a linear salt gradient whereupon arylsulfatase appeared as a single peak at 0.15 to 0.25 *M* NaCl. Gel filtration on Sephadex® G-200 also produced virtually identical elution profiles for arylsulfatase from the two species. By this procedure, the arylsulfatases from *A. sojae* SH 10-6 and *A. awamori* R 0827 were purified 60- and 45-fold with yields of 40 and 60%, respectively. Both enzymes were stable at pH 8.0 to 10.0 and were inactivated rapidly below pH 6.0, there being no apparent difference in this respect between the two enzymes. The activity of both enzymes was sustained fully for 1 hr at 20°C and pH 8.5, but prolonged incubation periods or higher temperatures caused gradual loss of activity.

b. *Properties*

Although Sakurai et al.[161] do not quote a molecular weight for the alkaline arylsulfatases, it is possible to estimate from the elution diagram for Sephadex® G-200 chromatography a value of about 60,000 for the enzyme from both organisms. This is in close agreement with the value of 57,000 reported by Burns and Wynn[157] for the corresponding enzyme (isoenzyme I) in *A. oryzae.* The pH curves for the *A. sojae* and *A. awamori* enzymes[161] are also very similar to each other, showing activity in the range 7 to 10 and maximum activity at pH 8.5. Thus, these enzymes are readily distinguished from all other *Aspergillus* isoenzymes so far described by their high alkaline pH optimum. They are also more susceptible to inhibition by certain inorganic ions than arylsulfatases II and III of *A. oryzae* (Table 6). Thus, Ag^+, Zn^{2+}, and Hg^{2+} were potent inhibitors of the alkaline arylsulfatases and Cd^{2+}, Al^{3+}, and Fe^{2+} were also inhibitory[161] whereas these ions (with the exceptions of Ag^+ which was not tested and Hg^{2+} which inhibited 50% of arylsulfatase II activity) generally produced little if any inhibition of arylsulfatases II and III of *A. oryzae.*[160] None of these cations was required as a cofactor for full activity, neither was ethylenediamine tetraacetate an inhibitor.

Cherayil's early study[153] had shown that despite differences in their relative susceptibilities (Table 7), arylsulfatases I, II, and III of *A. oryzae* were all inhibited by phosphate, fluoride, tungstate, and molybdate. However, Ohtsuki et al.[160] found that, although the alkaline arylsulfatases of *A. sojae* and *A. awamori* responded likewise to the presence of phosphate and fluoride, they were unaffected by tungstate and molybdate.

It is evident from their parallel behavior during purification and their similar properties that the alkaline arylsulfatases of *A. sojae* and *A. awamori* are very much alike. In fact, it would be difficult with current knowledge to distinguish between them. It is now equally apparent, however, that this close relationship does not extend to include arylsulfatase I of *A. oryzae* which differs significantly in certain properties (pH optimum, inhibition by cyanide, tungstate, and molybdate). As for the alkaline arylsulfatase activities in other *Aspergillus* species, nothing is known of their properties, but their widespread occurrence[160] suggests that the activity serves a specific and important purpose that none of these fungi can afford to lose. What that role is still remains to be discovered.

5. Arylsulfatases of A. nidulans

The survey by Harada and Spencer[147] of the ability of fungi to produce arylsulfatases included four species of *Aspergillus* among the 23 fungi found to be productive, namely *A. sydowi, A. oryzae, A. terreus* and *A. nidulans.* The last three of these ranked among the most prolific producers of arylsulfatases when grown on a standard medium containing taurine as sulfur source, but only *A. nidulans* produced significant activity when grown on a glucose-peptone-yeast extract medium. This fungus has attracted some attention as a source of arylsulfatase suitable for the study of the genetic and physiological regulation of the synthesis of arylsulfatase (see Volume II, Chapter 3.).

a. Purification

In the course of an investigation of the control of arylsulfatase synthesis in *A. nidulans,* Apte and Siddiqi[168] purified the two distinct fractions that were formed on derepression. Extracts of mycelia grown on a minimal dextrose medium were subjected to fractionation first by ammonium sulfate (60 to 100% saturation) and then by ethanol (40 to 70% v/v). After repetition of the ethanol treatment (which was effective in stabilizing the activity), the sample was treated with deoxyribonuclease and ribonuclease to degrade nucleic acids and finally chromatographed on DEAE-cellulose. The arylsulfatase activity eluted in two distinct fractions which were designated fractions I and II in order of elution in the salt gradient. These fractions represented purifications of 140- and 190-fold, with recoveries of 0.5 and 2.3%, respectively.

b. Properties

Among several aryl sulfates tested,[168] phenyl sulfate, *p*-nitrophenyl sulfate (K_m = 0.035 and 0.1 m*M* for fractions I and II, respectively) and nitrocatechol sulfate (K_m = 0.95 and 0.5 m*M*, respectively) were good substrates for both enzymes which also showed some activity toward phenolphthalein disulfate, but none toward indoxyl sulfate or choline sulfate. ^{18}O-Labeling studies[169] showed that hydrolysis of *p*-nitrophenyl sulfate occurs exclusively by cleavage of the O–S bond, in keeping with other arylsulfatase enzymes and also with nonenzymic cleavage of aryl sulfate esters (Table 1, Volume II, Chapter 2). Both fractions show maximum activity in a broad range of pH from 6.5 to 8.7 with a sharp fall in activity above pH 9.2. The effects of some inhibitors are shown in Table 7. Sulfite and cyanide are strong inhibitors, but phosphate and sulfate cause only weak inhibition. Fraction II is not susceptible to iodoacetic acid whereas Fraction I is inhibited 70%, reflecting a general rule that Fraction I is more sensitive to inhibitors than Fraction II.[168]

The activities of Fractions I and II show identical dependence on temperature. The optimum temperature was 45°C and the activation energy calculated from Arrhenius plots was 9 kcal/mol. Both enzymes were stable up to 55°C and inactivated at 70°C at the same initial rates although Fraction I enters a slower, second phase of inactivation, suggesting further heterogeneity. Finally, Fractions I and II have the same isoelectric point (pH 3.5 to 4.0) and molecular weight (44,000 calculated from sedimentation coefficients, Table 4).[168]

Clearly from the foregoing paragraphs, the two fractions of arylsulfatase in *A. nidulans* are very similar in most physicochemical characteristics: molecular weight, substrate specificity, activation energy, and isoelectric point. They are much more alike than any two of the three *A. oryzae* arylsulfatases. Substrate specificity and sensitivity to phosphate, sulfate, and cyanide allows both *A. nidulans* isoenzymes to be classified as Type 1. However, in certain aspects they do differ; they have slightly different affinities for substrates, different surface charges, and different sensitivities to certain inhibitors. Nevertheless, the question arises whether the two are entirely different proteins or whether one is a modified form of the other. The observed differences could be due to a configurational change or a chemical modification of a single protein. In this context it is interesting that during derepression of arylsulfatase synthesis in *A. nidulans*,[170] fraction II appears first followed by fraction I. This behavior would be consistent with a modification of fraction II to produce fraction I. It should be relatively straightforward to test this hypothesis by pulse-labeling with a labeled amino acid during growth, but unfortunately such experiments have not yet been performed.

6. Location of Arylsulfatases in Aspergillus Species

The very method whereby Taka Diastase® is prepared suggests that the enzymes which it contains are either secreted by *A. oryzae* mycelia into the surrounding medium or are loosely associated with the cell wall so as to be readily removed by the simple expedient of washing with water. Shiota's group[160,161] have also obtained extracts containing arylsulfatase from *A. oryzae* and other *Aspergillus* species by suspending mycelia in water or dilute buffer, and shaking. On the other hand, some workers have purified arylsulfatases from extracts prepared by mechanically or chemically disrupting mycelial pellets isolated from the growth medium.[158,168] Furthermore, although *A. oryzae* arylsulfatases can be separated from such extracts, a significant proportion of arylsulfatase remains associated with the particulate fraction even after homogenization and centrifugation.[162] On the basis of this indirect evidence alone, three compartments of arylsulfatase can be tentatively identified, namely an extracellular activity, and a mycelial activity consisting of soluble and particulate components. Reiss[171] has shown that in several fungi including *A. oryzae* and *A. nidulans*, arylsulfatase can be demonstrated histochemically with 6-bromonaphthyl 2-sulfate and Fast Garnet GBC as coupling agent. In fungal mycelia the strongest activity was detected in granules (probably lysosomes) in apices of the hyphae and the differentiating parts of conidiospores, in short in the most actively metabolizing parts. Presslitz has shown that all three of the arylsulfatase isoenzymes of *A. oryzae* can be found in the mycelium, but only two of them occur as extracellular enzymes.[156] Unfortunately no further details are available in published form. It would therefore seem that a reexamination of the distribution of arylsulfatases in *A. oryzae* is overdue, particularly now that the procedure of Burns and Wynn[162] for the independent assay of each isoenzyme is available. In addition, in view of the apparent extracellular location of part of the activity, it would be prudent in future investigations of the regulation of arylsulfatase synthesis to take due account of activity in the extracellular milieu. Discussion of the subcellular location of arylsulfatases in *Aspergillus* and other fungi in relation to their physiological roles can be found in Volume II, Chapter 4.

C. Arylsulfatases of *Neurospora crassa*

In the laboratories of Marzluf and Metzenberg, *N. crassa* has proved to be a useful model system for the study of the control of protein synthesis in eukaryotic cells (see Volume II, Chapter 3). One of the proteins chosen for study, primarily because of the extreme ease of its assay, was arylsulfatase. Consequently knowledge of arylsulfatase activity in terms of its relationship to the overall function and development of the cell is particularly rich for this organism. Unfortunately there is very little published work relating to the pure enzyme. In fact although two papers[172,173] have referred to purified arylsulfatase from *N. crassa* and a third[174] has indicated that DEAE-cellulose is useful for separating the arylsulfatase from choline sulfatase, no detailed method for the purification of arylsulfatase from this fungus has been published. Most attention has focused on the physiological and genetic regulation of the synthesis of the enzyme (see Volume II, Chapter 3). Related to this work are a number of peripheral studies on the general properties of the arylsulfatase activity, in particular those relating to the ability of the enzyme to contribute to the function and development of the cell (e.g., stability, subcellular location, developmental changes). It is these studies that are relevant to the present chapter.

1. Arylsulfatase in Conidia

During studies on the regulation of arylsulfatase synthesis in *N. crassa* Metzenberg and Scott[172,175] established a multiple location of the enzyme in conidia. When conidia were suspended in water or buffer, part of the activity was readily washed away, whereas the remainder (the "patent enzyme" so called because it was detected without rupturing the cells) remained associated with the cells. Successive washes or a long continuous wash did not release additional enzyme. When the washed conidia were treated with acid, part of the patent enzyme (20 to 37%) was inactivated although there was no loss of conidial viability. The remaining acid-resistant patent enzyme was not inactivated even after prolonged treatment. However, if the conidia were mechanically or chemically disrupted (see below), none of the arylsulfatase activity showed acid-resistance. Several trivial explanations for the existence of an acid-resistant patent enzyme were suggested and eliminated, namely that

1. Conidial proteins and polysaccharides buffer against added H^+ ions.
2. Synthesis of "new" arylsulfatase occurs during the 2 hr assay following acid treatment.
3. Arylsulfatase occurs as an acid-resistant proenzyme or in an internal compartment inaccessible to H^+.
4. Acid treatment generates an inhibitor rather than inactivates part of the enzyme.

Mechanical disruption of conidia or brief treatment with heat, chloroform, acetone, phenethyl alcohol, or polyene antibiotics resulted in a large increase in arylsulfatase activity in suspensions of conidia. At the same time, all of the enzyme activity became sensitive to acid. The amount of newly exposed enzyme (the "cryptic" enzyme) was independent of the method used to reveal it, but varied with each batch of conidia. Although all the treatments made the cryptic enzyme accessible to substrate, only pressure-cell disruption rendered it soluble. In all other treatments, enzyme was quantitatively retained on filter papers and rapidly sedimented by low speed centrifugation, suggesting an association of the enzyme with membranes of some form. The K_m of the cryptic enzyme was 2 mM for *p*-nitrophenylsulfate at pH 8.1, the same as those observed for patent acid-resistant enzyme and, apparently, for the purified enzyme. The pH-activity curve for the cryptic enzyme showed a sharp maximum at pH 7.6 to 7.8, again like the patent and the purified enzyme. However, despite similar K_m values and pH profiles, the weight of evidence was against the acid-resistant patent enzyme activity being identical with cryptic enzyme.[172]

Although the question was not completely resolved whether the acid-resistant patent enzyme is really in a separate compartment, the results seemed to be compatible with the existence of acid-resistant patent enzyme that is somehow completely accessible to protons but still protected from denaturation. In this context it may be relevant that in dormant conidia of *N. crassa*, disulfide groups are relatively abundant and they include oxidized glutathione (GSSG), glutathione-protein disulfide bonds and possibly protein-protein disulfide bonds.[176] It has been proposed that the high levels of disulfide bonds play an important role in maintaining a dormant state. Enzymes internally or externally cross-linked in this way might be relatively inactive and more stable, and account for the low metabolic activity and high stability of dormant conidia.[176] In the same vein, the patent acid-resistant enzyme could possibly be a more extensively cross-linked version of the acid-sensitive enzyme.

To summarize, *N. crassa* conidia contain at least three and probably four compartments of arylsulfatase. These are

1. A soluble exoenzyme.
2. A cell-bound fraction outside the plasma membrane and bound to the cell wall so that it is readily accessible to substrate (the patent enzyme).
3. An intracellular fraction (the cryptic enzyme).
4. Possibly a fraction bound to, or embedded in, the plasma membrane (the acid-resistant patent enzyme).

Whether these categories represent sublocations within each conidium or whether they represent subclasses of a heterogeneous population of cells is still unknown.[172] However, what is clear from counts of viable cells is that none of these categories arises from dead spores or mycelial debris.[172]

In many species, enzyme activities that are located in more than one compartment of a cell occur as structurally different isozymes, one for each location, but there is currently no evidence that this is the case for the conidial arylsulfatases. All of the enzyme is acid-sensitive in disrupted spores and the kinetic data collected so far reveals no significant differences (K_m for *p*-nitrophenyl sulfate = 2 m*M* and pH optimum = 7.6 to 7.8 for purified arylsulfatase, cryptic enzyme, and acid-resistant patent enzyme).[172] However, this is not sufficient cause to exclude the possibility that structural differences might still be found. For example, the invertase of yeast is polydisperse with respect to polysaccharide[177] and although the purified internal and external enzymes have the same K_m, pH optimum, molecular weight, and specific activity, they differ in that the external enzyme contains cystine or cysteine and about 50% mannan whereas the internal enzyme contains none of these.[178] In view of the additional observation that the mycelial arylsulfatase of *N. crassa* shows signs of heterogeneity (see the following), it would be unwise to assume that the enzymes in different compartments of conidia are all structurally identical.

An interesting feature of the multiple locations of arylsulfatase in conidia is the variability of distribution that occurs from batch to batch.[172,175] Thus the fraction of the total activity that occurs as the true extracellular enzyme varies between 30 and 75%, and the amount of patent activity in washed conidia that is resistant to acid varies from 20 to 37%. These changes probably reflect a redistribution of enzyme that occurs during conidial germination. Many enzymes in *N. crassa* enjoy an increased production during germination although some, notably the extracellular ones, decrease in amount, some change location, and different isoenzyme patterns can emerge.[176] Arylsulfatases in *N. crassa* reflect all these phenomena. During germination of conidia, the specific activity of the intracellular arylsulfatase activity increased dramatically, but very little was secreted into the medium[172,175] and the extracellular enzyme had almost completely disappeared at the mycelial stage of development. The increase in specific activity of the intracellular enzymes probably reflects the general activation of metabolism that is associated with the breaking of dormancy.[176]

2. Arylsulfatase in Mycelia

Samples of young mycelia exhibited what appeared to be a patent arylsulfatase activity, but the amount of enzyme detected depended on the buffer used for washing the cells.[175,179] In particular, cells washed in acetate buffer (pH 5.0) not only had more measurable activity than those washed in Tris buffer (pH 8.1) or succinate (pH 4.9), but they also lost nearly all viability. Chloroform washing produced similar effects to acetate, and acid treatment of Tris-washed cells led to loss of all arylsulfatase activity and viability within 10 min. The results suggested that in mycelia there is also a cryptic compartment of arylsulfatase, but the permeability barrier is much more fragile than in conidia. Such fragility complicates the determination of patent enzyme in mycelia if any exists at all.[172,175]

When mycelia of wild-type *N. crassa* were disrupted and centrifuged, a small but significant amount of arylsulfatase activity remained associated with the pellet fraction.[179] Experiments with a mutant unable to synthesize arylsulfatase indicated that the particulate location was probably not an absorption artifact. Some activity was lost whenever the particulate fraction from wild-type cells was washed, but some arylsulfatase always remained clearly associated with the particulate material. The arylsulfatase was readily solubilized by shearing forces or by the addition of a nonionic detergent, but the activity did not show latency. Both the lack of latency and the sedimentation properties of the particulate enzyme were unaffected by the method of disruption of mycelia. Just as fragility of mycelia complicates the assay of patent enzyme in whole cells, so the lability to shear of the intracellular particulate fraction complicates the interpretation of the observation that the arylsulfatase is not latent; even if the enzyme is membrane-enclosed, particles could lyse in the first few minutes of assay. The association of arylsulfatase and of acid phosphatase with the particulate fraction was roughly parallel and although certain differences suggested that each was associated with a different subpopulation,[179] it is probable that the particles were fungal lysosomes.[180-183] Scott[175] reported that arylsulfatase in *N. crassa* was contained in vesicles distributed irregularly throughout the mycelial hyphae. More recently, Reiss[171] has shown that most arylsulfatase is located in the growing tips of hyphae, side branches of the mycelium, and in the differentiating parts of conidiophores. The implications of these locations are considered elsewhere (Volume II, Chapter 4).

Arylsulfatase of *N. crassa* mycelia (wild-type and mutant strains) appears to be a resilient protein. It is relatively stable both in vivo,[184-186] and in vitro[187] at pH as low as 3.0. However, at pH 2.0 the activity was lost immediately even at 0°C.[187] The enzyme also succumbs to heat inactivation at 70°C with a half-life of 12 min at pH 8.0 and about 20 min at pH 7.5.[173,187] The decay was first-order except for the first 10 to 15% which was more rapid than the remainder.[173] This biphasic inactivation indicated possible microheterogeneity in the arylsulfatase which may be related to the multiplicity of enzymes observed in conidia. Gel electrophoresis of mycelial extracts lends some support to this suggestion because arylsulfatase stained with indoxyl sulfate occurred as wide bands and in some cases a faint slower band appeared for both wild-type and revertant mutant cells.[187] Possibly the slower band corresponds to small amounts of patent enzyme. Interestingly the invertases of yeast behave in this way with the internal enzyme having higher electrophoretic mobility than the external (glycoprotein) enzyme.[178] As Marzluf and Metzenberg[187] pointed out, the occurrence of two bands of arylsulfatase on gels does not constitute evidence for two enzymes of separate origins. In fact, the indications are that the enzyme is subject to a monomer-dimer equilibrium.[187]

The above indications that arylsulfatase from revertants and wild-type are the same was supported by kinetic data.[173] For all types studied, values of K_m for *p*-nitrophenylsulfate and nitrocatechol sulfate were 0.53 and 0.67 m*M*, respectively, and the activation energy for the catalytic hydrolysis was about 13 kcal/mol.[173]

3. Arylsulfatase in Ascospores

Ascospores of wild-type *N. crassa* do not contain arylsulfatase even when formed under the conditions of sulfur limitation that cause derepression of its synthesis in mature cells.[188] However, when the spores arise from a cross between the wild type and the constitutive mutant denoted *scon* ^{c} in which the synthesis of arylsulfatase is nonrepressible,[189] they do contain significant arylsulfatase activity even when they are formed in conditions of sulfur abundance.[188] About one half of the ascospores are colorless and inviable, although they are all morphologically normal.[176] Since only 1% of the spores that do germinate are *scon* ^{c}, the colorless inviable spores were presumed to be *scon* ^{c} progency. The failure of the trait of nonrepression to be transmitted with normal frequency in a cross indicated that untimely syntheses of enzymes of sulfur utilization (including arylsulfatase) might be lethal in some stage of meiosis or spore maturation. Indeed although conidia derived from the *scon* ^{c} mutant were morphologically normal, they appear to be osmotically fragile, and it has been suggested that perhaps the constitutive synthesis of these enzymes during conidiation results in formation of either a defective plasma membrane or an altered cell wall.[176] Whether the abnormal presence of arylsulfatase or related enzymes causes inviability of *scon* ^{c} ascospores, or whether it is simply another symptom resulting from the primary genetic defect in the *scon* ^{c} mutant, remains to be determined.

The debilitating effects that the untimely synthesis of arylsulfatase has in weakening the cell envelope, may be related to the more useful roles played by arylsulfatases and other lysosomal enzymes in the normal function of fungi. For example, release of ascospores[183] and conidiospores,[171] formation of anastomoses[183] (fision points of hyphae), and penetration of host cells by parasitic fungi[183,190,191] all require the cleavage of membranes and this is accomplished in all these cases by lysosomal enzymes. Further discussion of arylsulfatases in such processes is given in Volume II, Chapter 4.

D. Arylsulfatases from Other Fungi

Besides the sources of arylsulfatases that have so far featured in this chapter, a variety of other fungi have been shown to synthesize these enzymes. While only one of these has served as a starting point for purification, other studies have been made on the properties of enzymes in crude extracts or whole cells. These studies are now mentioned not only for the sake of completeness, but also because the contexts in which some of the work has been undertaken are of particular interest.

Several species of *Aspergillus* have already been mentioned that produce arylsulfatases. Early work by Harada[144] showed that *A. glaucus, A. usamii,* and *A. niger* produced considerable amounts of phenolphthalein and *p*-nitrophenol from the respective sulfate esters when the latter were present in the growth medium. *Rhizopus nigricans* was also particularly active in this respect, and, to lesser extents *Penicillium glaucum* and *Mucor rouxii.* The alkaline arylsulfatases of *A. oryzae, A. sojae,* and *A. awamori* examined in some detail by Shiota and others[160,161] (Section III.B.4) also occur in several other *Aspergillus* species including *A. usamii, A. inuii, A. saitoi,* and *A. niger* var. *fermentarious.*

Studies by Harada and Spencer[147] surveyed 27 fungi, of which only four *(Nematospora gossypii, Saccharomyces cerevisiae, S. cerevisiae* var. *ellipsoideus, Sporobolomyces salmonicolar)* were unable to produce arylsulfatase that could be detected in acetone-dried mycelia using *p*-nitrophenyl sulfate as substrate. Moreover, virtually no *p*-nitrophenol was liberated when *p*-nitrophenyl sulfate was present as sole sulfur source in the medium supporting growth of these four fungi. The other 23 fungi including species of *Rhizopus, Mucor, Neurospora, Ustilago, Aspergillus, Penicillium, Torulopsis,* and *Fusarium* all produced *p*-nitrophenol from its sulfate ester in the growth medium under appropriate conditions, and also produced arylsulfatase detectable in dried mycelia after growth on taurine as sole source of sulfur. Only two fungi *(Rhizopus stolonifer* and *Aspergillus nidulans)* produced detectable arylsulfatase following growth on glucose-peptone-yeast extract medium.

Interest in antibiotic production and how it relates to sulfur metabolism in *Cephalosporium acremonium* led Dennen and Carver[192] to study arylsulfatase production in this fungus. Following extraction from washed mycelia and subsequent concentration, the enzyme was partly purified (19-fold) by DEAE-cellulose column chromatography. Whole cells and partly purified arylsulfatase preparations had a K_m for *p*-nitrophenyl sulfate of 0.4 m*M* at pH 7.0 and 60°C (the enzyme being stable at 65°C for at least 30 min). The optimum pH was 5.5 in acetate buffer and 7.0 in Tris buffer. The molecular weight of the native enzyme was 50,000 and it always eluted from Sephadex® G-75 or DEAE-cellulose as a single peak. Cyanide and choline sulfate were inhibitory, but methionine and sulfate had no effect on the activity of the enzyme.

Milazzo and Lougheed[193] reported briefly upon the conditions required for synthesis of an arylsulfatase in the wood-rotting Basidiomycete, *Fomes geotropus.* Following this work, arylsulfatases have been detected in a number of other plant pathogens and, moreover, they have been implicated in the penetration of host cells by parasitic fungi. The plant pathogens *Ceratocystis fagacearum, C. fimbriata, Fomes annosus, Agaricus campestris, Alternaria tenuis, Piptocephalis virginiana, Mycotypha microspora,* and *Botrytis cinera* all possessed lysosomal particles that contained arylsulfatase.[182,183] As was the case with *N. crassa* the lysosomes were most abundant in hyphal tips.[183] During mycoparasitism in which one fungus is parasitic to another, arylsulfatase and deoxyribonuclease were concentrated at the surface of the haustorium (a modified hypha which invades the host cell and through which the parasite extracts nutrients). It thus appeared that lysosomal enzymes including arylsulfatase were actively involved in the process of parasitism.[183,190] Arylsulfatases have also been found in lysosome-like granules in *Rhodotorula rubra,*[194] *Saccharomyces cerevisiae* (despite an earlier failure to detect any activity in yeasts) and *Entomophthora coronata,*[195] an insect parasite that also causes a disease state (entomophthoromycosis) in man and other vertebrates.

Finally, dermatophytes are known to liberate a number of lytic enzymes into their immediate surroundings. The digestion of the surrounding milieu is a prerequisite for the survival of these skin parasites. Among the enzymes released from *Epidermophyta floccosum* is an arylsulfatase active toward *p*-nitrophenyl sulfate.[196] Primary isolates did not usually produce measurable arylsulfatase and all cultures remained arylsulfatase-negative in subsequent cultures until pleomorphic areas (mycelia) appeared in culture, at which time arylsulfatase was produced. In further subcultures, mycelia were rich in arylsulfatase while the brown granular conidia from the same initial isolates remained devoid of activity. The ability of *E. floccosum* to synthesize arylsulfatase thus appeared to be related to its morphology.[196] This behavior is not unlike the increase in intracellular activity observed during the development of *N. crassa* to the mycelial stage (see Section III.C.).

IV. BACTERIAL ARYLSULFATASES

A. Discovery

Shortly after Derrien's[117] discovery of arylsulfatase activity in mollusks, the first signs of the presence of a similar activity in microorganisms appeared in 1916 when Hoppe-Seyler[197] reported that urinary indican (indoxyl 3-sulphate) was decomposed by bacteria present in the urine. Hoppe-Seyler appeared to be unaware of the molluscan arylsulfatase and it is therefore interesting that studies on the indigo dyes and their precursors lead to the independent discovery of arylsulfatases on two separate occasions and in quite distinct organisms. Moreover a similar situation was to recur over

30 years later. These early discoveries, however, did little to stimulate interest in bacterial arylsulfatases and subsequent advances in the knowledge of the microbial enzymes was to come from studies not with bacterial sources, but with fungi, in particular extracts of *A. oryzae* (see Section III.B.). Apart from an isolated report of the presence of arylsulfatase in a *Mycobacterium* species,[146] it was not until the postwar years that bacterial arylsulfatases were rediscovered and a more persistent study made of them. At this time, Harada began an important series of studies that continue to this day and that have provided much of the current knowledge of the physiological and genetic regulation of synthesis of bacterial arylsulfatases (Volume II, Chapter 3).

In 1948, Harada reported the somewhat bizarre observation that "pieces of newspaper in a tub of feces and urine or in public latrines are often found to be dyed deep red."[198,199] The appearance of the stain in the paper required the presence of three factors: a bacterium in the urine, indoxyl sulfate in the urine, and lignin in the paper.[199,200] Without the lignin, a mixture of red and purple pigments was deposited in the growth medium, and these were identified as indirubin and indigo, respectively.[200] It was soon established that the common denominator that leads to the formation of these various pigments was the initial hydrolysis of indoxyl sulfate by a bacterial arylsulfatase.[201] Harada's curiosity concerning the red-stained newspaper led him to rediscover independently the phenomenon observed over 30 years earlier by Hoppe-Seyler.[197] Here then was the third occasion on which cleavage of indoxyl sulfate to initiate color-forming reactions was responsible for the discovery of an arylsulfatase enzyme. Harada and colleagues isolated[199] and eventually identified[202,203] the responsible bacterium as a new strain of *Streptococcus* which they called *S. alkalophilus* to indicate its tolerance of alkaline growth conditions.

At this time there was growing concern in some quarters over the inadequacy of microscopic appearance and colony morphology as a means of classifying various pathogenic bacteria and attention was increasingly turning toward methods of classification based on biochemical properties of microorganisms. The relative ease of detection of arylsulfatase activity made it an attractive component of this type of approach. At first, screening experiments of this kind were not very successful. For example, of 160 strains of *Staphylococcus pyogenes*, only two hydrolysed phenolphthalein disulfate present in the growth medium. More marked activity was observed in *Salmonella paratyphi* B and in an anaerobic spore-forming bacillus encountered as a contaminant.[204] A wider search covering 96 strains from eight genera (*Neisseria, Veillonella, Streptococcus, Staphylococcus, Diplococcus, Gaffkya, Sarcina*, and *Micrococcus*) revealed only two more bacteria capable of hydrolysing phenolphthalein disulfate.[205] However, in later studies, particularly with the mycobacteria,[206-208] arylsulfatase activity toward phenolphthalein disulfate proved to be a useful criterion for classifying different types of microorganisms.

Harada also examined the distribution of microbial arylsulfatases. Bacteria in urine and soils were positive in the lignin-coloration test, but bacteria found in stale fish meat and butter were negative.[198] Turning to pure stock cultures, Harada discovered that many microorganisms could substitute for *Streptococcus alkalophilus* in the lignin-coloration reaction.[209] Six of nine strains of *Aerobacter aerogenes*, a few undetermined nonspore-forming bacteria and two of twenty-two strains of *Escherichia coli* were able to hydrolyse indoxyl sulfate. However, it should be pointed out here that the apparent inability of some bacteria to produce arylsulfatase may reflect either inappropriate growth conditions or the unsuitability of certain arylsulfates as assay substrates, rather than the inability to synthesize the enzyme. The importance of exogenous nutrients in the regulation of arylsulfatase synthesis is dealt with in Volume II, Chapter 3. As for the second possibility, phenolphthalein disulfate is certainly not a universal substrate for microbial arylsulfatases. This was clearly shown when a variety of bacteria and fungi were compared for their ability to hydrolyse phenolphthalein disulfate, *p*-nitrophenyl sulfate and indoxyl sulfate present separately in the growth medium.[145] Three of the six strains of *A. aerogenes* tested showed strong activity towards *p*-nitrophenyl sulfate and two of these also hydrolysed indoxyl sulfate, but none was active toward phenolphthalein disulfate. Thirteen strains of *E. coli* were similarly examined and of these, five exhibited weak activity toward *p*-nitrophenyl sulfate or indoxyl sulfate and again none hydrolysed phenolphthalein disulfate. In the case of 14 strains of *Salmonella,* two *(S. paratyphi* A and *S. schottmuelleri)* were active, ten weakly active, and two inactive, irrespective of the substrate used. These results generally concurred with, and partly explained, those of Whitehead et al.[206,207] who earlier had used phenolphthalein disulfate as substrate and failed to detect any arylsulfatase in *Aerobacter, E. coli, Klebsiella,* or *Proteus.* However, like Harada, they did find some activity in *Salmonella* strains using this substrate, particularly in all 11 strains tested of *S. paratyphi* A and eight of eleven strains of *S. paratyphi* B *(Schottmuelleri).* Clearly the choice of substrate is important in screening experiments of this kind. Other striking examples also exist which illustrate the unsuitability of phenolphthalein disulfate as substrate. Thus, although Whitehead et al.[207] found no activity toward this compound in *Pseudomonas aeruginosa* and *Serratia marcescens* these organisms have since served as useful sources for the purification and characterization of arylsulfatases (see later sections).

Despite these clear indications, several reports have since appeared in the literature in which only phenolphthalein disulfate was used as substrate in screening tests for microbial sulfatases.[210-214] In fairness, it should be pointed out that in most of these cases the work has been concerned not with a search for arylsulfatase-producing organisms per se, but rather with establishing differences among organisms or groups of organisms. To this end, of course, it does not matter greatly which substrate is used so long as various types of microorganisms respond to it in different ways. Nevertheless, for those interested in the distribution of arylsulfatases in microorganisms, caution is necessary whenever a negative result for arylsulfatase activity is encountered, and due account should be taken of the number and type of arylsulfate esters that have been employed in the test. Indeed, if all the organisms that have so far failed to demonstrate arylsulfatase activity were to be reexamined with due attention to current knowledge of substrate specificity and factors regulating synthesis of the enzymes, it is likely that many, if not all, would yield a positive result.

Some bacterial arylsulfatases have been purified and characterized (Tables 8 and 9), but even taking into account the enzymes that have been examined in relatively crude extracts, the number so far studied in any detail is a small fraction of the total that are known to exist. The most extensively studied arylsulfatases have been obtained from organisms of the Enterobacteriaceae, Achromobacteriaceae and other families in the order of the Eubacteriales. Arylsulfatases from three *Pseudomonas* species have also attracted close attention. Finally, organisms of the genus *Mycobacterium* and related genera of the Actinomycetes have been widely screened for arylsulfatase activity although very little is known about the properties of the enzymes themselves.

B. Arylsulfatases of the Enterobacteriaceae

Most of the bacteria mentioned in the preceding section that rank among the earliest arylsulfatase-producers to be discovered belong to the family *Enterobacteriaceae.* Subsequent studies purifying and characterizing the bacterial arylsulfatases have continued to center on this family. The reason for such preoccupation probably stems not from an interest in their arylsulfatases per se, but from the fact that, because they are genetically well characterized organisms, they are suitable systems for the study of the regulation of arylsulfatase synthesis. Such investigations usually require that at some stage

Table 8
BACTERIAL AND ALGAL ARYLSULFATASE ISOENZYMES AND THEIR OPTIMUM pH VALUES

Organism	Isoenzymes	Optimum pH[a]	Ref.
Aerobacter aerogenes		7.1	222
Klebsiella aerogenes		7.5	224,226
Salmonella typhimurium	5 isoenzymes (electrofocusing)	6.7[b]	231,233
Serratia marcescens		6.8—8.2	234
Proteus vulgaris		6.1; 5.8 with NCS	235
P. rettgeri	Fractions I—IV (DEAE-cellulose). II, III, and IV further resolved by paper electrophoresis into cathodic (C) and anodic (A) forms		236,238, 239
	I C_I	8.6; 6.2—6.7 with NCS	
	II C_{IIa}	Assayed at 6.7 with NCS	
	II C_{IIb}	Assayed at 6.7 with NCS	
	II C_{IIc}	Assayed at 6.7 with NCS	
	III A_{III}	Assayed at 8.3 with NPS and 6.7 with NCS	
	III C_{IIIa}	Assayed at 6.7 with NCS	
	III C_{IIIb}	Assayed at 8.3 with NPS	
	IV[c] A_{IV}	Assayed at 8.3 with NPS and 6.7 with NCS	
	IV[c] C_{IV}	Assayed at 6.7 with NCS	
Alcaligenes metalcaligenes	—	8.75; 8.0 with NCS	244
Streptococcus alcalophilus	—	9.5—9.7	203
Pseudomonas aeruginosa	α	9.8; 8.8 with NCS	255
	β	8.4	
Pseudomonas C12B	—	8.8; 7.5—8.0 with NCS	260,261
Comamonas terrigena	Inducible arylsulfatase	Assayed at 7.5	264
	Tyrosine sulfate sulfatase	Assayed at 7.5 with TOS	265,266
Mycobacterium fortuitum	4 isoenzymes, using BNS	6.3[b], using MUS	284,287
Chlamydomonas reinhardti	—	9.8; 9.0 in presence of imidazole	288

[a] Using NPS as substrate unless stated otherwise. Abbreviations: NPS = *p*-nitrophenyl sulfate, NCS = nitrocatechol sulfate, TOS = tyrosine-*O*-sulfate, BNS = 6-benzoyl 2-naphthyl sulfate, MUS = 4-methylumbelliferyl sulfate.

[b] Mixture of isoenzymes.

[c] Gel electrophoresis of IV indicated 3 bands, the first active toward NPS alone, the second toward NCS alone and the third toward both compounds.

the arylsulfatase be purified and characterized so that factors that affect the activity of preformed enzyme, such as substrate specificity and the response to activators or inhibitors, can be distinguished from factors that regulate synthesis of the enzyme. As a result, a considerable volume of knowledge has accumulated on the properties of arylsulfatases in strains of *Klebsiella, Enterobacter* (formerly *Aerobacter), Salmonella, Serratia,* and *Proteus.* These members of the Enterobacteriaceae have many phenotypic properties in common suggesting that they constitute a single evolutionary group of bacteria. Indeed, Steffen et al.[215] have recommended that strains once assigned to *Aerobacter* and *Enterobacter* should all be reassigned to the genus *Klebsiella.* Furthermore, they suggest[216] that the Enterobacteriaceae can be divided into subgroup A, containing *Escherichia, Klebsiella (Aerobacter* and *Enterobacter),* and *Citrobacter;* and subgroup B containing *Serratia, Proteus* and *Erwinia.* Interestingly, Harada's group[217] have fairly recently reported a comparative immunological study on the aryl-

Table 9
EFFECT OF ANIONS ON BACTERIAL ARYLSULFATASES

Source of enzyme	PO_4^{3-}	SO_4^{2-}	CN^-	Cl^-	F^-	SO_3^{2-}	S_3^{2-}	Ac[a]	N_3^-	EDTA[b]	Ref.
Aerobacter aerogenes	−	O	=	O	−	O	·	·	O	O	218, 221
Klebsiella aerogenes	−	O	=	·	O	−	·	·	·	·	244
Salmonella typhimurium	−	·	·	·	·	·	·	·	·	·	231,233
Serratia marcescens	−	O	−	·	=	=	·	·	·	·	234
Proteus vulgaris	=	−	O	+	=	=	·	+	·	·	235
P. rettgeri[c]											
C_I	±[d]	O	−	O	·	·	·	O	·	·	236—239
C_{IIa}	O	−	O	O	·	·	·	O	·	·	
C_{IIb}	O	O	O	O	·	·	·	−	·	·	
C_{IIc}	O	O	O	−	·	·	·	O	·	·	
A_{III}	[e]	O	−	O		...	O	·	·		
C_{IIIa}	O	O	O	O	·	·	·	−	·	·	
C_{IIIb}	O	O	−	O	·	·	·	O	·	·	
A_{IV}	[e]	−	=	O	·	·	·	[f]	·	·	
C_{IV}	O	O	O	O	·	·	·	O	·	·	
Alcanigenes metalcaligenes	O	O	=	O	O	−	·	·	·	·	244
Pseudomonas aeruginosa α	O	O	−	O	·	·	·	O	·	·	255
P. aeruginosa β	+	+	−	O	·	·	·	O	·	·	
Pseudomonas C12B	·	O	·	O	·	·	·	·	·	·	259—261
Comamonas terrigena	·	O	·	·	·	+	−	·	·	·	267

Note: +, Weak to moderate activation; O, no effect; −, weak to moderate inhibition; =, strong inhibition; •, not determined.

[a] Acetate.
[b] Ethylenediamine tetraacetate.
[c] See Section IV.B.5.b and Table 8 for description of notation of isoenzymes.
[d] Inhibition with *p*-nitrophenyl sulfate (NPS); activation with nitrocatechol sulfate (NCS).
[e] Inhibition with NPS; no effect with NCS.
[f] Inhibition with NCS; no effect with NPS.

sulfatases of members of the Enterobacteriaceae and arrived at an almost identical classification based on cross-reactivity of bacterial proteins to antiserum raised against the pure arylsulfatase of *K. aerogenes* W70. The specific antiserum formed precipitin lines with protein from *E. coli, C. freundii, Salmonella typhimurium,* and seven strains of *Klebsiella* (six of which had originally been classed as *Aerobacter* or *Enterobacter)* but not *Serratia marcescens,* seven strains of *Proteus* or two strains of *Erwinia,* irrespective of whether or not each particular strain possessed an active arylsulfatase. Evidently the organisms in the first group produce a protein that is immunologically identical, either in whole or in part, with the *K. aerogenes* arylsulfatase, while the second group does not. Moreover, as will shortly emerge, arylsulfatases within each group possess some common properties. It therefore seems sensible to abide by this segregation in the following account.

1. Arylsulfatase of Aerobacter (Enterobacter) aerogenes

Although Whitehead et al.[207] had failed to detect arylsulfatase activity toward phenolphthalein disulfate in *A. aerogenes,* Harada and colleagues[145,218] showed that this was due to the use of an unsuitable substrate and that there is activity toward other aryl sulfates. In fact the discovery of arylsulfatase in *A. aerogenes* was a milestone in the study of sulfatases because it was during studies with this organism that the first important signs were found of the involvement of tyramine in the regulation of arylsulfatase synthesis.[219,220] Early preliminary studies[218] on the properties of the partially purified enzyme from *A. aerogenes* were followed in 1963 by a full-scale purification of the enzyme from the same source.[221]

a. Purification

The preliminary work of Harada[218] employed an extract of acetone-dried cells partly purified by ammonium sulfate precipitation (0 to 67%). Later the pure enzyme from cells grown on a sulfur-deficient medium was obtained by a combination of French press extraction, precipitation of nucleic acids with streptomycin sulfate, ethanol precipitation, ammonium sulfate fractionation, DEAE-cellulose, and TEAE-cellulose ion-exchange chromatography. The final preparation which represented a 350-fold purification in a 10% yield was homogeneous according to several criteria.[221]

b. Properties

Work with the pure enzyme confirmed some of the properties observed for the impure enzyme (e.g., optimum pH of 7.1, inactivity toward phenolphthalein disulfate, inhibition by cyanide and phosphate, but not sulfate). The pure enzyme hydrolysed *p*-nitrophenyl sulfate (K_m = 3.3 mM) and phenyl sulfate, but not naphthyl sulfate. It was inhibited by fluoride and by Zn^{2+} and Pb^{2+}. With these properties the enzyme was assignable[221] neither to Type 1 nor Type 2 classifications established for other arylsulfatase enzymes.

The arylsulfatase of *A. aerogenes* showed an unusual temperature-activity profile.[221] Between 0° and 28°C, a normal dependence of activity on temperature was observed and an Arrhenius plot gave an activation energy for catalytic hydrolysis of 11.3 kcal/mol. In the range 30 to 60°C, the rate was independent of temperature. The enzyme lost activity only slowly at temperatures below 60°C, but between 60 and 65°C the enzyme was denatured rapidly. The decrease in activity was so abrupt that this parameter was considered to be a valuable criterion for identifying the protein and in particular for distinguishing "wild type" from genetically altered arylsulfatase enzymes. The enzyme was also inactivated at pH values outside the range 6 to 8, and in the presence of high concentrations of urea.

Several lines of evidence pointed to the presence in *A. aerogenes* of a single arylsulfatase with a molecular weight of 40,700.[221] However, the enzyme slowly lost activity during cold storage and this loss was accompanied by the appearance of several additional protein peaks during ion-exchange chromatography. Although none of these peaks showed arylsulfatase activity, it is possible that this phenomenon might be related to the occurrence of multiple arylsulfatases in other bacterial extracts.

Finally, the enzyme was not secreted into the medium during growth, but sonic disruption released all the activity even with gentle treatment when no obvious disintegration of the cells had occurred.[222] The implication is that the arylsulfatase is cell-wall associated.

2. Arylsulfatase of Klebsiella aerogenes

Because the *A. aerogenes* strain employed by Harada and colleagues did not possess a suitable genetic exchange system, these workers were forced to turn their attention to other organisms to pursue their study of the regulation of arylsulfatase synthesis. The organism chosen was *Klebsiella aerogenes* W70 which is so closely related to *A. aerogenes* that they may be considered as two strains of the same species.[215] A survey of the desulfating ability of *K. aerogenes* W70 indicated that the bacterium produced one arylsulfatase active toward *p*-nitrophenyl sulfate, tyrosine-*O*-sulfate, and phenyl sulfate, and another enzyme specific for *p*-carboxyphenyl sulfate.[223] In the latter case, although significant amounts of sulfate were released no *p*-hydroxybenzoic acid was detectable in the incubation mixture using either cell suspensions or cell-free extracts. There was, therefore, some uncertainty as to whether the responsible enzyme, which was present irrespective of growth conditions, was a simple arylsulfatase. The other enzyme appeared during growth on medium containing either methionine (nonrepressed) or sulfate plus tyramine (derepressed). Purification and characterization of the arylsulfatases from nonrepressed and derepressed cells have shown that the same enzyme is produced in both sets of circumstances.[224]

a. Purification

Cells grown under nonrepressing or derepressing conditions were disrupted by sonication and the arylsulfatase purified from cell-free extracts by ammonium sulfate fractionation (55 to 80% saturation), and a sequence of column chromatography steps (DEAE-cellulose, hydroxylapatite, Sephadex® G-100 and DEAE-Sephadex®). A single arylsulfatase was present in the DEAE cellulose eluates (cf. *A. aerogenes)* indicating the presence of a single arylsulfatase in this organism. The overall purification was 1400-fold with yields around 7%, and pure enzyme from nonrepressed and derepressed cells had the same specific activity, same migration on polyacrylamide gels, and the same molecular weights.[224] More recently this lengthy procedure has been superseded by simple and very effective, affinity chromatography methods.[225] The affinity medium consisted of tyrosine coupled to beads of β-1,3-glucan through a hexamethylenediamine spacer. When used in conjunction with the first three steps of the previous purification method (extraction, ammonium sulfate fractionation, and DEAE-cellulose chromatography) a purification of 1000-fold was achieved with yields of around 50% and sometimes as high as 85%. Despite the slightly lower purification factor compared with the original method the product was electrophoretically homogeneous. Alternatively, the tyrosine affinity medium could be replaced by an immunoadsorbant carrying antibody for the purified *K. aerogenes* W70 arylsulfatase.[228] Specific antibody from rabbits was coupled to the same β-1,3 glucan matrix. Speed, convenience, good recoveries, and high purifications achieved by these methods recommend then for wider application in the study of arylsulfatases.

b. Properties

The *K. aerogenes* W70 arylsulfatase is a monomeric protein with a molecular weight between 45,000 (gel-filtration) and 47,000 (sodium dodecyl sulfate-gel electrophoresis) that is located in the periplasmic space.[224,226] The optimum pH is 7.5 and nitrophenyl sulfates (*o*-, *m*-, and *p*-isomers) and nitrocatechol sulfate are the best substrates, although several other arylsulfates are also hydrolysed (Table 8). Generally speaking, affinity of enzyme for substrate and maximum velocity of hydrolysis were both enhanced by the presence of electron-withdrawing substituents.[224] The mechanistic implications of this behavior which is also found for other bacterial arylsulfatases (for example, *Alcaligenes metalcaligenes*) is discussed in Volume II, Chapter 2. Phosphate, cyanide, and hydroxylamine were inhibitory, but not sulfate or fluoride. The inhibitor and substrate specificity patterns thus permit assignment of the *Klebsiella* arylsulfatase to neither Type 1 nor Type 2 classifications. A variety of cations were without effect although NH_4^+ ions caused activation. In most of these properties, the enzyme is very similar to the *A. aerogenes* arylsulfatase emphasizing the close taxonomic relationship between these bacteria.

The pure arylsulfatase from *K. aerogenes* W70 was also inhibited by tyramine and the structurally related compounds, dopamine and octopamine, but not by tryptamine or histamine.[224] This inhibition contrasts with the tyramine-mediated activation of other arylsulfatases (e.g., enzymes from *A. oryzae*, Section III.B.3, and *Pseudomonas* C12B, Section IV.E.2). Moreover, the specificity of inhibition is intriguing, particularly in view of the specific involvement of tyramine in the regulation of synthesis of the enzyme (Volume II, Chapter 3).

Finally, mention should be made of the fact that although *K. aerogenes* W70 was originally considered to produce only a single arylsulfatase, indications have recently come to light[4] that some additional active protein bands appeared during purification under certain conditions. Whether this reflects production of multiple forms of the enzyme by the cell or artifacts of the purification procedure, is not clear.

3. Arylsulfatase of Salmonella typhimurium

Despite earlier reports in the literature that certain strains of *S. typhosa*[227,228] and *S. typhimurium*[145] produced arylsulfatase after prolonged growth on a complex medium, there have been several recent claims that the latter bacterium does not synthesize arylsulfatase.[217,229,230] However, Henderson and Milazzo[231] found that if tyramine was included in the growth medium, not only was the arylsulfatase readily detectable, but there was also sufficient present to allow partial purification.

a. Purification

S. typhimurium strain LT2 grown to late log phase on minimal medium containing tyramine produced high levels of arylsulfatase. The cells were acetone-dried and extracted into Tris buffer. Arylsulfatase activity was separated by ammonium sulfate fractionation (30 to 90% saturation) and further purified by Sephadex® G-200 gel filtration chromatography. Enzyme activity was confined to a single peak that represented a purification of about 20-fold over the crude extract.

b. Properties

The partly purified enzyme, which was monomeric with a molecular weight of 53,500 to 55,700, migrated as a single band on conventional polyacrylamide gel electrophoresis.[231] However, the higher resolving power of isolectric focussing revealed five isoenzyme bands with isoelectric points at pH 6.0, 6.2, 6.3, 6.4, and 6.6. Charge differences among the isoenzymes of mammalian arylsulfatase A had previously been shown to be due to variation in sialic acid contents.[232] Consequently, the *S. typhimurium* LT2 isoenzymes were treated with neuraminidase prior to electrofocussing to see if they too could be accounted for in this way.[233] However, the isoenzyme pattern remained unchanged and it was concluded that the different forms do not differ in hydrolysable sialic acid content. The occurrence of multiple arylsulfatases is not uncommon in microorganisms and a general discussion of the phenomenon is included at the end of this chapter.

The arylsulfatase activity of *S. typhimurium* LT2 (i.e., the five isoenzymes taken collectively[231]) hydrolyses *p*-nitrophenyl sulfate (K_m = 7.9 m*M*) and nitrocatechol sulfate (K_m = 4.1 m*M*) at the pH optimum of 6.6 to 6.7. Relatively high activity toward nitrocatechol sulfate together with observed inhibition by phosphate is indicative of a Type 2 enzyme, but until response to cyanide and sulfate are also known, this assignment must be considered tentative.

The enzyme is very similar in certain respects to other enzymes in subgroup A of the *Enterobacteriaceae*. For example, the arylsulfatase activity from *S. typhimurium* LT2, like the *K. aerogenes* W70 enzyme, is inhibited by tyramine, an intriguing observation in view of the ability of tyramine to stimulate arylsulfatase synthesis in both bacteria. The inhibition of the *S. typhimurium* LT2 enzyme is competitive, with inhibition constants of 4.5 m*M* with *p*-nitrophenyl sulfate as substrate and 7 m*M* with nitrocatechol sulfate.[233] In addition, the enzyme is able to withstand temperatures up to 55°C for 1 hr without appreciable loss of activity. Raising the temperature to 75°C results in marked decrease in activity within 5 min.[233] This behavior is reminiscent of the arylsulfatase from *A. aerogenes*. Arylsulfatases from all three bacteria have molecular weights around 50,000, maximum activity near neutral pH, and are inhibited by phosphate.

4. Arylsulfatase of Serratia marcescens

Having segregated the *Enterobacteriaceae* according to cross-reactivity with antiserum for arylsulfatase of *K. aerogenes* W70, Harada and co-workers[234] proceeded to examine the arylsulfatase in *S. marcescens* which, although known to produce an arylsulfatase, was one of the bacteria devoid of the antigenic determinants present in the *Klebsiella* enzyme.[217] Four strains of *S. marcescens*, and one each of *S. indica* and *S. marinorubra* were tested, and in all strains arylsulfatase synthesis was repressed by sulfate or methionine and, according to the authors (see Volume II, Chapter 3), derepressed by addition of tyramine. *S. marcescens* IFO 3046 produced the highest levels of activity and this organism was used for further experiments.[234]

a. Purification

Stationary phase cells of *S. marcescens* IFO 3046 grown on a mannitol-sulfate medium and derepressed by the addition of tyramine were disrupted by sonication and the arylsulfatase separated from the cell-free extract by ammonium sulfate fractionation (45 to 65% saturation). The crude enzyme was then purified by a combination of DEAE-cellulose, CM-cellulose and DEAE-Sephadex® column chromatography steps. The final preparation represented a purification of 1250-fold and 21% recovery. Polyacrylamide gel electrophoresis of the purified enzyme indicated purity in excess of 90%.

b. Properties

The electrophoretic mobility of the *S. marcescens* IFO 3046 arylsulfatase under nondenaturating conditions is less than that of the *K. aerogenes* enzyme, although their molecular weights are remarkably similar (Table 4). The *S. marcescens* enzyme resembles arylsulfatases of the subgroup A organisms in several other ways.[234] For example, *p*-nitrophenyl sulfate is the best substrate for enzymes from *S. marcescens, K. aerogenes,* and *A. aerogenes;* phosphate, cyanide, and hydroxylamine inhibit all three; tyramine inhibits the arylsulfatases of *S. marcescens*, *K. aerogenes*, and *S. typhimurium*. On the other hand, there are also some distinct differences: electrophoretic mobilities differ; fluoride inhibits the arylsulfatases of *A. aerogenes and S. marcescens,* but not *K. aerogenes*; *o*- and *m*-nitrophenyl sulfates and nitrocatechol sulfate are almost as good as *p*-nitrophenyl sulfate as substrates for the *K. aerogenes* arylsulfatase, but are relatively inferior when used with the *S. marcescens* enzyme. Clearly, although the *S. marcescens* arylsulfatase does not contain antigenic determinants in common with the enzymes of subgroup A organisms and although they show different responses to certain substrates and inhibitors, they also have several common features. Thus, there may be an evolutionary relationship between the *S. marcescens* enzyme and those of subgroup A bacteria, but if so it is probably a remote one. Whether or not there is such a relationship must await more direct comparisons of, for example, amino acid sequences of the individual enzymes.

5. Arylsulfatases in the Genus Proteus

This genus was classified[217] alongside *Serratia* in subgroup B (Section IV.B.) as having no antigenic determinants in common with the arylsulfatase of *K. aerogenes* W70. In contrast with *S. marcescens*, however, the available information concerning the properties of the *Proteus* arylsulfatases more strongly emphasizes the dissimilarity between subgroups A and B.

Two species of *Proteus* have attracted attention as sources of arylsulfatases. Arylsulfatase in *P. vulgaris* (NCTC 4636) was described in an early paper by Dodgson[235] at a time when the Type 1-Type 2 classification had just been established. The few microbial arylsulfatases that had been extensively investigated had all been assigned to the Type 1 class and the *P. vulgaris* arylsulfatase was the first example of a microbial Type 2 enzyme. The arylsulfatase of *P. rettgeri* was later described in a series of papers by Milazzo and Fitzgerald,[236,239] and this study was particularly notable for revealing extensive heterogeneity in the arylsulfatase content of this organism. Although multiple arylsulfatases had already been detected in mammals and fungi (see Section III), this was the first time they had been demonstrated in bacteria. As mentioned elsewhere, more examples of multiple forms of bacterial arylsulfatases have since come to light (e.g., *S. typhimurium*, Section IV.B.3), and this feature is discussed at the end of this chapter.

a. Arylsulfatase of P. vulgaris

No attempts have been made to purify arylsulfatase from this species. However, studies[235] with extracts of acetone-dried cells previously grown on nutrient broth have provided some basic information and revealed certain features that clearly relate the enzyme to Type 2 arylsulfatases in higher organisms (see Table 9).

Using *p*-nitrophenyl sulfate and nitrocatechol sulfate, maximum activity was observed at pH 6.1 and 5.8, respectively, independent of the substrate concentration, and K_m values at the respective pH optima were 55 and 5 m*M*. The greater affinity for nitrocatechol sulfate and the strong inhibition by phosphate, fluoride (and to a lesser extent, sulfate), but not cyanide, identified the enzyme as Type 2. Furthermore, phenylhydrazine and semicarbazide are without effect on the enzyme (neither do they inhibit Type 2 enzymes of the mollusks *Patella vulgata* and *Helix pomatia*). In contrast, these reagents are powerful inhibitors of Type 1 enzymes (see, for examples, sections on *A. metalcaligenes* and *A. oryzae*, this chapter). The inhibitory effects of some anions appeared to be pH-dependent. Thus, the strong inhibition of activity toward *p*-nitrophenyl sulfate by phosphate, fluoride, and sulfite occurred most predominantly on the acid side of the optimum pH with the result that these ions produced a shift in apparent optimum pH to higher values simultaneously with an overall diminution of activity. Fluoride had a parallel effect on activity towards nitrocatechol sulfate, and other Type 2 enzymes (e.g., liver arylsulfatase B) show similar behavior.[240-242]

On the other hand, chloride activated the *P. vulgaris* enzyme. With nitrocatechol sulfate as substrate, there was a modest activation only on the acid side of the pH optimum so that a downward shift in optimum pH was apparent. With *p*-nitrophenyl sulfate, the enzyme activity doubled throughout the range of pH used and therefore no shift was observed. Activity toward both substrates was also enhanced by raising the concentration of acetate used to buffer the assay medium. Kinetic measurements indicated that both chloride and acetate influenced the breakdown rather than the formation of the enzyme substrate complex.

The interactions underlying the effects of anions on the arylsulfatase of *P. vulgaris* are still not understood. Nevertheless these observations clearly reinforce the assertion that the arylsulfatase of *P. vulgaris* is a Type 2 enzyme, and quite distinct from the arylsulfatases of *K. aerogenes, A. aerogenes*, and *S. typhimurium*.

b. Arylsulfatase of P. rettgeri

Intact cells of *P. rettgeri* grown on defined medium containing methionine as sole source of sulfur hydrolysed both nitrocatechol sulfate and *p*-nitrophenyl sulfate.[236] The effect of various parameters such as substrate concentration, pH, temperature, and the concentration of various ions were examined and some of the data, particularly the widely different pH optima for nitrocatechol sulfate (pH 6.7) and *p*-nitrophenyl sulfate (pH 8.3), suggested the presence of more than one arylsulfatase. Confirmation came in a subsequent paper from the same laboratory which described the separation of a total of nine forms of the enzyme by a combination of chromatography and electrophoresis.[238] Arylsulfatases in streptomycin-treated extracts were first separated into four component (I to IV) by chromatography on DEAE-cellulose. Fractions I and II contained activity toward nitrocatechol sulfate but not *p*-nitrophenyl sulfate, whereas fractions III and IV hydrolysed both esters. Each chromatographic fraction was then examined for further heterogeneity by paper electrophoresis,[238] polyacrylamide gel electrophoresis,[239] and density gradient centrifugation.[239]

Fraction I arylsulfatase migrated as a single band during density gradient centrifugation and during electrophoresis on paper and in polyacrylamide gels.[238,239] Although fraction I was originally considered to be specific for nitrocatechol sulfate,[238] later work showed that *p*-nitrophenyl sulfate could also serve as substrate.[239] The K_m values for nitrocatechol sulfate and *p*-nitrophenyl sulfate were 6 and 0.75 m*M* at the respective pH optima of 6.1 and 8.6. Cyanide, which strongly inhibited nitrocatechol sulfate hydrolysis, caused the optimum pH for the reaction to shift to 6.7. This shift is much like that produced by fluoride acting on the *P. vulgaris* enzyme, and these observations emphasize the need to study the influence of anions over a range of pH values. Chloride which caused activation and an acid-shift of the pH optimum of the *P. vulgaris* arylsulfatase, was without effect on the fraction I activity of *P. rettgeri*, but phosphate inhibited hydrolysis of *p*-nitrophenyl sulfate and activated hydrolysis of nitrocatechol sulfate. The capacity for phosphate activation is shared only with one of the isoenzymes of *P. aeruginosa* (see Section IV.E.1.). It is worth pointing out that the initial stimulus for searching for multiple arylsulfatases in *P. rettgeri*, namely widely different pH optimum for the two substrates, still persists with this fraction. This, together with other differences (effects of phosphate and cyanide), suggests further heterogeneity although this was not apparent on polyacrylamide gels or during density gradient centrifugation.[239]

Fraction II arylsulfatase remained as a single band at the origin during attempts to fractionate by polyacrylamide gel electrophoresis.[239] However, heterogeneity in terms of size and shape was evident after density gradient centrifugation which produced two bands of arylsulfatase activity, both hydrolysing nitrocatechol sulfate but not *p*-nitrophenyl sulfate. This same fraction II was resolved into three forms by paper electrophoresis and although these all had the same substrate specificity and were unaffected by cyanide or phosphate, they were distinguishable by different responses to inhibition by sulfate, chloride, and acetate (Table 9).

Fraction III, like fractions I and II, was unresolved by gel electrophoresis at pH 7.4. However, at pH 8.5 two bands appeared, one hydrolysing nitrocatechol sulfate but not *p*-nitrophenyl sulfate, and the other hydrolysing both compounds. Density gradient centrifugation produced only a single band. Paper electrophoresis revealed three bands. Two of these moved towards the cathode and were specific for *p*-nitrophenyl sulfate and nitrocatechol sulfate. A third band moved toward the anode and was active toward both substrates. These forms were further distinguished by different responses to anionic inhibitors.

Fraction IV, when subjected to paper electrophoresis at pH 7.4, produced one anodic band active toward *p*-nitrophenyl sulfate and nitrocatechol sulfate, and a cathodic form active toward nitrocatechol sulfate alone, the latter being particularly notable in being the only one of the nine isoenzymes that was unaffected by cyanide, sulfate, phosphate, chloride, or acetate.[238] Polyacrylamide gel electrophoresis of this fraction at pH 7.4 also produced two bands and although one of these was an anodic band active toward both substrates, the other was specific, not for nitrocatechol sulfate, but for *p*-nitrophenyl sulfate.[239] This apparent contradiction in specificities was resolved by gel electrophoresis at pH 8.5 which revealed three anodic bands, one of which was active toward *p*-nitrophenyl sulfate alone, one toward nitrocatechol sulfate alone, and one active toward both substrates.[239] An important feature of the fraction IV subforms is the occurrence of an isoenzyme specific for *p*-nitrophenyl sulfate. Of the nine subforms in *P. rettgeri,* only one other (present in fraction III) was found with this specificity.

It is clear that the isoenzyme pattern of arylsulfatase in *P. rettgeri* is complex, and that this complexity is compounded by the variety of responses that the isoenzymes show toward inorganic anions. Purification of individual isoenzymes and comparison of structures, catalytic, and other properties is now badly needed in order to bring some degree of unity to what at present is a somewhat bewildering picture. Meanwhile, it is very difficult to draw any further comparisons with arylsulfatases of other bacteria other than to mention that this pattern of multiple forms is the most complex yet encountered.

C. Arylsulfatases of the Genus *Alcaligenes*

Considerable knowledge of the properties of the arylsulfatases in the *Enterobacteriaceae* has accumulated over the last 20 years or so. However, even before this period, studies in the Cardiff laboratories had revealed in some detail properties of the enzyme from *A. metalcaligenes*. The organism was isolated from intertidal mud in the Bristol Channel.[243] At the time of the work, the science of enzyme purification was in a far less refined state than it is today. Nevertheless, the level of purification achieved was sufficiently high to allow detailed studies on some properties of the enzyme.

1. Arylsulfatase of A. metalcaligenes

a. Purification

Cells grown on nutrient broth under aerobic conditions were harvested, acetone-dried and extracted into water. Arylsulfatase was precipitated from the extracts by acetone precipitation (85% v/v) and dried *in vacuo*.[243] This preparation was used for some preliminary studies of the properties of the enzyme. Subsequently, attempts were made to achieve further purification by ammonium sulfate fractionation, adsorption on anion exchangers, alumina, calcium carbonate, and benzoic acid, but these approaches all resulted in considerable losses of activity and were therefore of little value.[244] However, some success was obtained with paper electrophoresis at pH 8. The final preparation which was free of esterase, phosphatase, chondrosulfatase, glycosulfatase, and glucuronidase, represented a 300-fold purification with recovery of 35% of the arylsulfatase activity in crude extracts.[244] Dodgson and co-workers used this preparation to reveal many of the properties to be described. In more recent years, the methods has been modified[245] by replacing the paper electrophoresis step by gel filtration on Sephadex® G-100, DEAE-cellulose, and isoelectric focusing.

b. Properties

Initial studies[243] with crude arylsulfatase prepared by acetone precipitation showed that *p*-acetylphenyl sulfate and *p*-nitrophenyl sulfate could serve equally well as substrates. Optimum substrate concentrations were 3 and 1.5 m*M*, respectively, the optimum pH with either compound was 8.75 and in both cases the enzyme was inhibited by excess substrate. Using nitrocatechol sulfate, however, no inhibition by excess substrate was observed and maximum activity was measured at a 15 m*M* concentration. With this substrate the optimum pH shifted to 8.0. Such anomalous behavior of nitrocatechol sulfate is discussed further in Section VI. The relatively alkaline pH optimum of the enzyme which is similar to that of certain arylsulfatase isoenzymes found in fungi (Table 5) is particularly interesting in relation to the pH of the estuarine mud (8.1 to 8.2) from which the organism was isolated and to the ability of the organism to produce alkali.

Using the arylsulfatase purified by paper electrophoresis, similar pH and substrate curves were obtained.[244] Interestingly, although the V_{max} value for nitrocatechol sulfate was about 15 times less than those of *p*-nitrophenyl sulfate and *p*-acetylphenyl sulfate, the former substrate gave the lowest K_m (0.22 m*M*, compared with 0.475 and 0.9 m*M*, respectively). In a subsequent paper,[246] the range of aryl sulfates able to serve as substrates was extended to include, on the one hand, those containing electron-withdrawing groups (e.g., *o*-, *m*-, and *p*-nitro groups, *o*- and *p*-chloro groups) and on the other hand, those containing electron-donating substituents (e.g., *o*-, *m*-, and *p*-amino-methyl, and -hydroxy groups). In general, the K_m values decreased and V_{max} increased as electron-withdrawal into the aryl moiety became more extensive.[246] More recently a similar behavior was observed with arylsulfatase from *K. aerogenes*.[224] Both *p*-nitrophenyl sulfate and *p*-acetylphenyl sulfate were hydrolysed at the O–S bond[247] by the *A. metalcaligenes* arylsulfatase (Table 1, Volume II, Chapter 2). In relation to foregoing comments concerning the naturally occurring aryl sulfate esters (Section II), it is particularly interesting that tyrosine-*O*-sulfate is a very good substrate for the *A. metalcaligenes* enzyme, being second only to *p*-nitrophenyl sulfate and about six times better than nitrocatechol sulfate.[248]

The enzyme is not affected by phosphate, sulfate, chloride, or fluoride, but it is inhibited by sulfite and cyanide.[244] These properties and the relatively low activity toward nitrocatechol sulfate led[142] to the classification of this enzyme as Type 1. Strong inhibition by cyanide was matched by comparable inhibition by phenylhydrazine, hydrazine, semicarbazide, thiosemicarbazide, and hydroxylamine, compounds generally regarded as ketone- or aldehyde-specific reagents.[244] The kinetics of inhibition by cyanide and hydroxylamine were of the anticompetitive type,[249] indicating that the binding of inhibitor followed the binding of the substrate. This requirement for the presence of substrate was confirmed by Shaw et al.[245] who found that irreversible binding of [^{14}C]-cyanide to arylsulfatase partly purified from *A. metalcaligenes* was most noticeable in the presence of substrate.

The pH dependence of K_m and V_{max} for *p*-nitrophenyl sulfate, *p*-acetylphenyl sulfate, and nitrocatechol sulfate showed the presence of two ionizable groups (pK_a 8.2 and 9.4) in the enzyme that participate in substrate binding.[244] Although the identity of the former was obscure, the latter was tentatively identified as α or ω-amino, phenolic OH, or thiol.

The preference for substrates containing electron-withdrawing substituents, the inhibition by cyanide and other "aldehyde" reagents, and the involvement of ionizing groups in substrate binding, all have a bearing on the mechanism of action of the enzyme and are therefore discussed further in Volume II, Chapter 2.

2. Arylsulfatases of Other Species of Alcaligenes

At the time of the isolation and identification of *A. metalcaligenes* from Bristol Channel mud, other species of *Alcaligenes* were examined for production of arylsulfatase activity during growth.[243] Of one strain each of *A. bookeri, A. haemolysans*, and *A. subviscosum*, seven of *A. faecalis*, and five unidentified species of *Alcaligenes*, only one *(A. faecalis* NCTC 1347) showed enzyme activity toward phenolphthalein disulfate. However, the growth medium contained nutrient broth and under these conditions it is common for arylsulfatase synthesis to be repressed (Volume II, see Chapter 3). In addition, phenolphthalein recoveries from most growing cultures were poor and it is possible that arylsulfatase activity was masked by degradation of phenolphthalein as it was liberated. Interestingly, *A. faecalis* 1347 which did exhibit arylsulfatase activity, did not destroy any phenolphthalein during growth. In view of these complications and the limitations imposed by using only phenolphthalein disulfate as potential substrate, it is clear that the other species of *Alcaligenes* that so far appear to be unproductive may well elicit arylsulfatase activity, given the right conditions.

D. Arylsulfatases in Other Members of the Eubacteriales

The abundance of organisms in the family Enterobacteriaceae that demonstrably produce arylsulfatase is matched by a dearth of such organisms in the remaining families of the Eubacteriales. This is not to say, however, that searches have not been made. On the contrary, members of the Micrococcaceae, Neisseriaceae, Lactobacillaceae, Corynebacteriaceae, and Bacillaceae have all been examined at one time or another for ability to produce arylsulfatase activity. It must be emphasized here that these searches were performed over 25 years ago. They therefore predate both the discovery of a rigid and complex regulation of arylsulfatase synthesis and the work on the substrate specificity of isolated enzymes. Consequently, the frequent failure to observe arylsulfatase activity in cultures of these organisms might simply reflect inappropriate choices of growth conditions and/or aryl sulfate substrates, rather than a genetic deficiency. A negative result for arylsulfatase is therefore always of questionable significance and this should be borne in mind in the following paragraphs.

Of 160 strains of *Staphylococcus pyogenes* and 75 coagulase-negative staphylococci, only two of the former and none of the latter liberated phenolphthalein from its disulfate ester during a 7- to 14-day incubation period.[204] Schultz-Haudt and Scherp[210] also had difficulty in demonstrating arylsulfatase activity in pure staphylococcal cultures despite the fact that they used a more appropriate substrate (phenyl sulfate). Various *Streptococci*, diphtheroid bacilli, microaerophilic Gram-negative cocci, and fusiform bacilli were equally unproductive. However, arylsulfatase activity was already known[250] to be present in human saliva, and the observation that 20 out of 21 aerobic cultures of mixed gingival bacteria produced activity when growing in a phenyl sulfate-broth medium suggested a symbiotic production of the enzyme.[210] Tests with binary combinations of isolated colonies revealed that a fusiform bacillus was indispensable if arylsulfatase was to be produced. It was suggested[210] that the fusiform bacillus produced the arylsulfatase and that the ancillary bacteria provided a low oxidation-reduction potential and accessory factors required by the fusiform bacillus for growth and arylsulfatase production. However, it is by no means clearly established which organism was producing the stimulating environment and which (if indeed only one of them) was synthesizing the arylsulfatase.

By this time, Harada had already demonstrated the involvement of bacterial arylsulfatase in the production of lignin red (Section IV.A.), and in 1954 published evidence was given for classification of the responsible bacterium as a *Streptococcus* species.[203] This particular species, *S. alcalophilus*, was characterized by its ability to grow in alkaline conditions, an ability reflected by alkaline pH optima of some of its enzymes including arylsulfatase (pH 9.5 to 9.7).[145,203]

More recently, arylsulfatase activity has been considered as a possible component of diagnostic tests to distinguish *Neisseria gonorrhoeae* and *N. meningitidis* from other species of *Neisseria, Branhamella,* and *Moraxella.*[251] However, the final selection of ten enzymes (from a total of 48) did not include arylsulfatase. Presumably the arylsulfatase test did not discriminate sufficiently, but whether this was because there was a preponderance of positive or of negative results is not clear.

Apart from reports showing arylsulfatase activity in two unidentified nonsporeforming bacteria,[209] and very weak activity toward indoxyl sulfate and *p*-nitrophenyl sulfate in some strains of *E. coli*,[145] the rest of the literature generally records negative results for other members of the Eubacteriales. These are not listed here for the reason presented earlier that negative results for arylsulfatase tests are of questionable significance. Indeed, this is amply illustrated by the fact that some organisms listed in early work[207] as nonproducers of arylsulfatase (such as *Proteus vulgaris, Pseudomonas aeruginosa,* and *Serratia marcescens)* have since served as sources for purification of arylsulfatase.

E. Arylsulfatases in the Genus *Pseudomonas*

The same three members of the *Pseudomonadaceae* that have served as sources in the study of alkylsulfatase enzymes (Volume I, Chapter 2), have also attracted attention for their ability to produce arylsulfatase activity, namely *P. aeruginosa, Pseudomonas* C12B, and *Comamonas terrigena.*

1. Arylsulfatase of P. aeruginosa

In 1952 Whitehead and colleagues[207] reported that *P. aeruginosa* grown in broth containing phenolphthalein disulfate or naphthyl 1-sulfate produced no arylsulfatase activity. However, during the following decade the repression of arylsulfatase synthesis by sulfate, cysteine, and other readily assimilatable sulfur compounds became known and this enabled Harada[252] to reexamine the problem in a new light. When the organism was grown in a citrate medium containing separately sulfate, sulfite, thiosulfate, or cysteine as sole sources of sulfur no arylsulfatase activity was detected.[253] When the sulfur source was either methionine or an alkyl sulfate ester (choline sulfate, serine-*O*-sulfate, or threonine-*O*-sulfate) an arylsulfatase was produced active toward phenyl sulfate, *p*-nitrophenyl sulfate, phenolphthalein disulfate, and *p*-carboxyphenyl sulfate.[253] However, if an aryl sulfate ester (phenyl sulfate, tyrosine-*O*-sulfate, or tyramine-*O*-sulfate) was the sulfur source, this enzyme was accompanied by an additional one that was able to hydrolyse tyrosine-*O*-sulfate and tyramine-*O*-sulfate. The possibility that cells grown with an aryl sulfate as sulfur source have only one arylsulfatase with a slightly modified specificity to that synthesized using other sulfur sources, was considered unlikely. It thus appeared that *P. aeruginosa* was able to synthesize at least two arylsulfatases.[253]

Further heterogeneity then emerged when Delisle and Milazzo[254] examined extracts of cells that, because they were grown on methionine as sulfur source, contained only one of the two activities observed by Harada. Arylsulfatase activity in an acetone powder of *P. aeruginosa* was purified and resolved by polyacrylamide gel electrophoresis into two distinct bands labeled I and II (and, later,[255] α and β). The fact that two enzymes were also found in cell extracts obtained by sonic oscillation was considered to eliminate the possibility that the two bands arose as artifacts of the purification procedure (unless, of course, sonication produced the same artifacts).

Of the three activities so far detected, the substrate-inducible enzyme of *P. aeruginosa* has not been examined further, but the α and β isoenzymes have been purified and characterized.[255]

a. Purification

P. aeruginosa was grown first on a citrate medium containing $MgSO_4$ and for a further period on a citrate-methionine medium to derepress arylsulfatase synthesis. Enzymes were extracted from acetone-dried cells and purified by DEAE-cellulose, DEAE-Sephadex®, ammonium sulfate fractionation and preparative gel electrophoresis.[254,255] This procedure produced an overall purification of more than 100-fold with recovery of 45% of the total arylsulfatase activity.[255]

b. Properties

The α and β isoenzymes have the same molecular weight (60,000), but differ slightly in charge, having isoelectric points of 4.85 and 4.90, respectively.[254] Using nitrocatechol sulfate as substrate, K_m values for the α and β enzymes were 2.04×10^{-4} *M* and 1.66×10^{-4} *M*, respectively, and with *p*-nitrophenyl sulfate 2.94×10^{-5} *M* and 7.5×10^{-5} *M*.[255] Evidently *p*-nitrophenyl sulfate is the preferred substrate for both isoenzymes although in excess it does produce inhibition which does not occur with nitrocatechol sulfate.

With either substrate, the optimum pH was 8.8 to 9.0 for the α enzyme and 8.4 for the β, although the latter value shifted toward pH 9.0 when *p*-nitrophenyl sulfate was used at concentrations exceeding the optimum.[255] This feature has been observed[242] with arylsulfatases from the glands and secretions of the snail *Helix pomatia*. The effect of pH on K_m and V_{max} for the α and β isoenzymes indicated the presence of ionizable substrate-binding groups with pK_a values for the α isoenzyme of 7.8 and 8.8 and, for the β isoenzyme, of 7.8 and 8.4. Amino, phenolic, or sulfhydryl residues were considered as possible identities although the last seemed unlikely as the enzyme was resilient to thiol reagents. The pK_a values are close to those of the *A. metalcaligenes* arylsulfatase and may indicate common mechanistic features (see Volume II, Chapter 2).

Cyanide inhibits both α and β arylsulfatases, but sulfate and phosphate do not.[255] These properties and the preference for *p*-nitrophenyl sulfate as substrate demonstrate the Type 1 nature of both isoenzymes. In fact, sulfate and phosphate activate the β enzyme when *p*-nitrophenyl sulfate is the substrate although chloride and acetate, which are known to activate arylsulfatases of *Proteus* (Section IV.B.5.), had no effect.[255]

Both isoenzymes were stable at pH 7 to 9 and in the 30 to 40°C temperature range, but rapidly lost activity at higher temperatures. Arrhenius plots were linear up to 40°C giving activation energies of 18 to 20 kcal in keeping with other microbial arylsulfatases. The discontinuity observed at 40°C is also found for other bacterial arylsulfatases (for example, *A. metalcaligenes*,[244] *A. aerogenes*[221]) and indeed for other enzyme systems for which it has been suggested that temperature-dependent conformational changes are involved.[256,258] These observations coupled with the substrate-dependent shift in optimum pH for the *P. aeruginosa* enzyme lead Delisle and Milazzo[255] to imply that conformational changes in arylsulfatases might be an aspect worthy of more attention. Unfortunately this area still remains neglected.

2. Arylsulfatase of *Pseudomonas* C12B

Fitzgerald and Payne[259] found that the regulation of synthesis of certain alkylsulfatases in *Pseudomonas* C12B was unusual when compared with the regulation of synthesis of arylsulfatase and choline sulfatase in other microorganisms. Consequently they investigated the ability of *Pseudomonas* C12B to synthesize arylsulfatase to see whether the regulation was characteristic of the bacterium or of the sulfatase type. This bacterium did indeed form an arylsulfatase, the synthesis of which was controlled in essentially the same way as in other bacteria and fungi (see Volume II, Chapter 3). More recently the enzyme, which was active towards nitrocatechol sulfate and *p*-nitrophenyl sulfate, has been purified and characterized more fully,[260] although part of the work is currently only available in thesis form.[261]

a. Purification

Pseudomonas C12B was derepressed for arylsulfatase synthesis by growth on L-methionine as sole source of sulfur. Crude extracts were then subjected to ammonium sulfate fractionation (60 to 90% saturation) prior to removal of nucleic acid by precipitation with streptomycin sulfate. DEAE-cellulose chromatography of the residual protein solution produced a peak of arylsulfatase activity that represented a 60-fold purification and contained four proteins separable by gel electrophoresis, only one of which possessed arylsulfatase activity.[260] Thus, *Pseudomonas* C12B, unlike other bacteria, does not possess multiple forms of the enzyme. The partial purification obtained by this published procedure can be completed by further combinations of column chromatography steps (Sephadex® G-150, Bio-Rex 70®, DEAE-cellulose, and DEAE-Sephadex®) according to a procedure developed in Fitzgerald's laboratory.[261] The final preparation which was homogeneous according to various criteria (polyacrylamide gel electrophoresis, immunodiffusion, immunoelectrophoresis), represented a 200-fold purification with recovery of 7.7% of the arylsulfatase activity. The low recovery arose mainly from the considerable denaturation that occurred throughout purification and it was improved by omitting certain chromatographic steps that contributed little to the purification, but led to appreciable losses of activity (e.g., Sephadex® G-150 chromatography).

b. Properties

The purified arylsulfatase of *Pseudomonas* C12B is a monomeric protein[261] with a molecular weight of about 51,500 (Table 4). An interesting feature of this enzyme is the stimulation of its catalytic action by the addition of tyramine in high concentrations[260,261] (above 10 m*M*). Maximum activation (about twofold) was produced at 40 m*M* tyramine and the same effect was observed both with crude extracts and with partly or completely purified enzyme. The ability of tyramine to stimulate was also independent of the culture conditions used to generate the enzyme, and of pH in the range 5 to 8 although the effect was enhanced at extremes of pH. Conversely the presence of tyramine did not affect the pH optimum of the enzyme which was 8.8 with *p*-nitrophenyl sulfate and about 7.5 to 8.0 with nitrocatechol sulfate.[260,261]

Activation by tyramine had been observed previously for one of the *A. oryzae* isoenzymes (Section III.B.3.b.) and was accounted for on the basis that the particular isoenzyme in question acted as a sulfotransferase, using tyramine or one of a number of other phenols as sulfate acceptors. However, the arylsulfatase of *Pseudomonas* C12B did not possess sulfotransferase activity; sulfate and phenol were released in equal amounts and all attempts to detect the formation of tyramine-*O*-sulfate were unsuccessful.[262] The possibility that tyramine was binding to the enzyme and modifying the active site or activating a second site was also eliminated by equilibrium dialysis experiments that demonstrated no binding between enzyme and tyramine.[261,262]

The vital clue to the mystery lay in the observation that the amount of tyramine required for maximum activation depended on the substrate concentration.[261,262] As the substrate concentration increased, the amount of tyramine required for maximum activity also increased indicating that the enhancement might be due to an interaction between substrate and tyramine. This was indeed the case. A complex between tyramine and *p*-nitrophenyl sulfate was detected after paper chromatography and moreover, arylsulfatase was not stimulated when tyrosine-*O*-sulfate was the substrate and this ester did not form a complex with tyramine.[261,262] Other arylamines including tryptamine, dopamine, and octopamine could substitute for tyramine. The K_m and V_{max} for *p*-nitrophenyl sulfate alone were 2.34×10^{-5} *M* and 5.5×10^{-3} μmol/min and in the presence of excess tryptamine 8×10^{-5} *M* and 8.52×10^{-3} μmol/min. Evidently the enzyme has less affinity for the tryptamine salt but hydrolysed it faster than the dissociated K^+ salt of the ester.[262] Although the underlying causes of tyramine-mediated activation of arylsulfatases in *Pseudomonas* C12B and *A. oryzae* are different, the same practical implications apply (see Section III.B.3.b.). Chloride, which activates the *P. vulgaris* arylsulfatase, and sulfate and cysteine had no effect on the *Pseudomonas* C12B enzyme.[259,260]

The substrate specificity of the *Pseudomonas* C12B enzyme is quite broad and includes *p*-nitrophenyl sulfate, nitrocatechol sulfate, *p*-acetylphenyl sulfate, 4-hydroxy 3-nitrophenyl sulfate, and tyrosine-*O*-sulfate, but not ascorbic acid 2-sulfate.[261]

The enzyme was released almost completely from intact cells after osmotic shock or after treatment with lysozyme suggesting that it was probably located in the periplasmic space of the cell wall.[263]

3. Arylsulfatases and Tyrosine Sulfate Sulfatases of Comamonas terrigena

This organism was coisolated with *Pseudomonas* C12B and is of particular note as being the first bacterium for which an inducible mechanism of arylsulfatase synthesis was discovered[264] (see Volume II, Chapter 3). Growth of *C. terrigena* at the expense of *p*-nitrophenyl sulfate as sulfur source occurred after a 35-hr lag period and was accompanied by the release of *p*-nitrophenol into the growth medium and by increased ability of cell extracts to hydrolyse *p*-nitrophenyl sulfate in vitro. When nitrocatechol sulfate or *p*-acetylphenyl sulfate was used as inducer, cell extracts hydrolysed nitrocatechol sulfate, *p*-acetylphenyl sulfate, *p*-nitrophenyl sulfate and tyrosine-*O*-sulfate at low but roughly comparable, rates at the assay pH of 7.5. With *p*-nitrophenyl sulfate as inducer, the activity increased markedly toward *p*-nitrophenyl sulfate, but not toward the other esters. It thus appeared that this compound was inducing the synthesis of an additional enzyme specific for its own hydrolysis. The cell extracts liberated sulfate and *p*-nitrophenol in roughly equimolar amounts, confirming the presence of a sulfatase activity. When tyrosine-*O*-sulfate was used as inducer, activity toward tyrosine-*O*-sulfate was the highest in relative terms although the activities toward all substrates were low. Again it appeared that tyrosine-*O*-sulfate was inducing a specific tyrosine sulfate sulfatase.

Subsequent work on arylsulfatase in this organism has concentrated on the inducible tyrosine sulfate sulfatase. Aged cell-extracts released sulfate and tyrosine from tyrosine-*O*-sulfate in a 1:1 ratio. However, with fresh cell extracts or intact cells, the rate of appearance of sulfate was lower than that of its coproduct, indicating that in these situations liberated sulfate was being used for biosynthesis.[265,266] The enzyme was essentially unaffected by the presence of 1 m*M* sulfate or tyrosine, inhibited by thiosulfate, and activated by cysteine or sulfite.[267] Although other microbial arylsulfatases are activated by certain anions (for example, enzymes of *P. aeruginosa, P. vulgaris,* and *P. rettgeri)* activation by cysteine or sulfite is unusual, particularly in the case of sulfite which has been found to be a strong inhibitor of some microbial arylsulfatases (Table 7 and 9).

Hydrolysis of tyrosine-*O*-sulfate by resting cells indicated a location for the responsible enzyme somewhere near the cell-surface. Cells subjected to osmotic shock or to treatment with sucrose-lysozyme-EDTA were relieved of most of their tyrosine sulfate sulfatase suggesting a cell-wall location.[265] However, the failure of sucrose alone to release more than trace amounts of activity, the total retention of activity by the cells during induction, and the requirement for EDTA to facilitate penetration of peptidoglycan by lysozyme indicated that the enzyme was not at the periphery of the cell. The implied periplasmic location for the tyrosine sulfate sulfatase was further supported by the release of most of the activity from osmotically shocked cells.[265]

Interestingly, resting cells of *C. terrigena* also synthesized a constitutive tyrosine sulfate sulfatase that was found in extracts of cells irrespective of prior exposure to tyrosine-*O*-sulfate.[265] This enzyme was virtually undetectable in intact cell suspensions and was retained by the cells after osmotic shock and after EDTA-lysozyme treatment. Evidently this constitutive tyrosine sulfate sulfatase was not cell-wall associated.[265]

When extracts of cells were examined using polyacrylamide gel electrophoresis,[266] those induced with tyrosine-*O*-sulfate produced two enzyme bands that released tyrosine and sulfate from tyrosine-*O*-sulfate. Extracts of cells not exposed to tyrosine-*O*-sulfate contained only the slower moving enzyme. This enzyme was also the only form of arylsulfatase in *C. terrigena* grown with cysteine or methionine as sole source of sulfur, and it should therefore be considered as a constitutive form of tyrosine sulfate sulfatase until such time as evidence for endogenous production of tyrosine-*O*-sulfate is forthcoming.

Both the inducible and constitutive forms of tyrosine sulfate sulfatase were distinct from a third arylsulfatase present in tyrosine-*O*-sulfate-induced and noninduced bacteria.[266] This band which was observed on gels stained with nitrocatechol sulfate, was close to, but reproducibly distinct from, the inducible tyrosine sulfate sulfatase, and may correspond with the small amounts of arylsulfatase produced during growth in the presence of aryl sulfates other than *p*-nitrophenyl sulfate or tyrosine-*O*-sulfate.[264]

In short *C. terrigena*, like many other microorganisms, produces a multiplicity of enzymes with hydrolytic activity toward aryl sulfates, including one inducible and one constitutive tyrosine sulfate sulfatase, and one or possibly two arylsulfatases of broader specificity. The relationship among these arylsulfatases in terms of their molecular and catalytic properties and their collective physiological role remains unknown.

F. Arylsulfatases in the Genus *Mycobacterium* and Related Genera

The mycobacteria are characterized by filamentous or mycelial growth and some of them are responsible for producing well-known pathogenic conditions in man such as pulmonary tuberculosis *(Mycobacterium tuberculosis)* and leprosy *(M. leprae)*. Indeed, it is concern for the identification, control, and prevention of such conditions that is indirectly responsible for the considerable literature on the distribution of arylsulfatase activity in this genus. Not all mycobacteria are pathogenic for man and an important factor in the control, containment, and treatment of these disease states is the ability to classify and identify the mycobacteria as particular species. Because testing for arylsulfatase activity is so easy using chromogenic aryl sulfate esters, it has enjoyed considerable popularity as one of a battery of biochemical tests used in the identification of these bacteria.

According to Harada,[146] arylsulfatase activity was observed in a *Mycobacterium* species as early as 1943. However, not until 1951 did more detailed studies on the distribution of arylsulfatase among the mycobacteria become available when Whitehead and colleagues[207,208] reported the presence of arylsulfatase in various bacteria of medical interest. Initially ten strains were examined by growing each in broth containing phenolphthalein disulfate and observing the release of the parent phenol. Of four strains isolated from cold-blooded animals, two strains of *M. piscium* and one of *M. ranae* showed considerable activity while a strain of *M. chelonei* was less active. Of the remaining saprophytic mycobacteria, two strains each of *M. phlei, M. smegmatis,* and *M. stercoris* were tested, and one of each pair gave a negative result and the rest were weakly positive. This initial work, suggesting a widespread distribution of arylsulfatase in the mycobacteria but with different degrees of activity, led to the examination of a larger number of strains for ability to hydrolyse aryl sulfates present in the culture medium.[208] All strains tested (one avian, one from a cold blooded animal, four bovine, and four human strains of *M. tuberculosis,* one strain each of *M. phlei,* and *M. stercoris,* and seven strains of saprophytic mycobacteria) liberated 1-naphthol from naphthyl 1-sulfate present in the growth medium during a 14-day test period. When naphthyl sulfate was replaced by phenolphthalein disulfate, the *M. tuberculosis* strains were not active whereas most of the others exhibited some activity.

At this point in the investigation attention was confined exclusively to phenolphthalein disulfate because it was obviously the more useful substrate for discriminating between certain groups of mycobacteria. Consequently many mycobacteria have been described in the literature as nonproducers of arylsulfatase activity whereas in fact the observed failure to liberate phenolphthalein is more likely due to the unsuitability of phenolphthalein disulfate as a substrate. A similar situation has already been described for the Eubacteriales (Section IV.A). Moreover, Whitehead et al.[208] showed that five bovine strains and one human strain of *M. tuberculosis* could degrade phenolphthalein, thus masking any arylsulfatase activity that might be present.

Although this turn of events has been detrimental to the accumulation of knowledge of mycobacterial arylsulfatases themselves, it has nevertheless been of great use clinically in identifying and classifying pathogenic and nonpathogenic mycobacteria. Whitehead et al.[208] proceeded to examine 154 strains of this genus. None of 66 human strains of *M. tuberculosis* and only 3 of 27 bovine strains were able to hydrolyse phenolphthalein disulfate (1 m*M* concentration) in a 14-day test, but most avian and murine strains of this species and most saprophytic mycobacteria were able to do so. Wayne et al.[211] attempted to extend this approach to the problem of distinguishing between saprophytic mycobacteria and the tubercle bacilli (i.e., human or bovine strains of *M. tuberculosis)* and were particularly concerned to find some way to include certain "atypical" mycobacteria and strains of *Nocardia* in the scheme. However, the tests proved variable and no useful information emerged. Interestingly, almost all species of mycobacteria tested hydrolysed demonstrable amounts of phenolphthalein disulfate if sufficient inoculum was used and allowance made for permeability differences. However, only 20% of the *Nocardia* strains, and none of the *Streptomyces* or *Corynebacterium* strains produced activity toward this substrate.

Further studies of this kind emerged over the next few years[212,214,268-270] culminating in a comprehensive paper[271] describing the identification of mycobacteria on the basis of a variety of tests including arylsulfatase activity. According to this scheme the mycobacteria were first divided into slow and rapid growers. Organisms within each group were distinguished by combinations of about 30 characteristics including hydrolytic activity toward phenolphthalein disulfate. Of the slow growers (which include the tubercle bacilli) none exhibited any activity during a 3-day incubation. However, after 14 days while the tubercle bacilli remained unproductive, 80% of the other strains *(M. kansasii, M. marinum, M. terrae, M. intracellulare, M. scrofulaceum,* and *M. xenopi*) had liberated phenolphthalein from its disulfate ester. For the fast growers, the 3-day test proved positive for 94% of the *M. fortuitum* strains tested, thereby distinguishing them from other rapidly growing saprophytes *(M. phlei, M. chitae, M. aurum, M. parafortuitum,* and *M. smegmatis)* that are not usually clinically significant.

Other studies delineating the biochemical characteristics of the Actinomycetes continued to reveal further examples of arylsulfatase-producing organisms, including some species of *Nocardia,*[272,273] pathogenic strains of *M. xenopi,*[274] *M. borstelense* isolated from a chronic abscess,[275] *M. abscessus* from an infected puncture wound,[276] and others.[277,278] Advancing knowledge led to further classification schemes[279,280] that included an arylsulfatase test, and this is now incorporated routinely in procedures for identifying the mycobacteria.[281]

Despite the obvious widespread occurrence of arylsulfatases in the genus *Mycobacterium* almost nothing is known of the properties of the enzymes themselves. In their early study, Whitehead et al.[207] prepared cell-free extracts of *M. piscium* grown in the presence or absence of phenolphthalein disulfate and found that maximum arylsulfatase activity toward this ester was obtained at pH 6.3 with a substrate concentration of 2 to 3 m*M*. Examination of the original data suggests a K_m of about 0.5 m*M*. Since that time many more mycobacteria have been shown to hydrolyse phenolphthalein disulfate and one of the most prolific producers of arylsulfatase is *M. fortuitum*.[282,283]

The *M. fortuitum* complex is a heterogenous group of rapidly growing acid-fast bacilli consisting of *M. chelonei* (subspecies *chelonei* and *abscessus), M. fortuitum,* and *M. peregrinum*, and these potentially pathogenic species are the only rapid growers that produce arylsulfatase in large quantities.[284] In order to clarify the taxonomic status of the complex and the relationship among its members, Binko and Bartholomew[284] examined arylsulfatase multiplicity in cell-free extracts of each organism by electrophoretic and immunological methods. Using agarose electrophoresis and staining for activity with 6-benzoyl 2-naphthyl sulfate,[285] arylsulfatase zymograms were obtained for extracts of *M. chelonei* (eight strains), *M. fortuitum* (four strains), and *M. peregrinum* (two strains). Each produced two bands, all of which moved anodically (Figure 8). Extracts of *M. chelonei* contained the fastest (S_5) and the slowest (S_1) migrating bands thereby distinguishing this organism from the other two. The faster bands (S_4) of *M. fortuitum* and *M. peregrinum* were coincident, but these species were distinguished by the relative mobilities of their slower bands (S_3 and S_2, respectively). The slower band (S_3) in strains of *M. fortuitum* migrated more rapidly, and stained more heavily, than S_2 of *M. peregrinum*. Zymograms were the same for all strains of a given species, irrespective of culture age or protein concentration. By turning to polyacrylamide gel electrophoresis, the resolution of bands was improved considerably and each of the fourteen strains tested conformed to one of four distinct patterns (Figure 8). Patterns I and II were produced by strains of *M. chelonei*; all strains of this organism possessed bands C_1 and C_3, and two strains of *M. chelonei* subsp. *chelonei* produced a third band, C_2, of intermediate mobility. The *M. fortuitum* zymograms contained a weakly staining, but highly mobile band F_1, well separated from a diffusely staining zone that was further resolved by extending the electrophoresis to reveal three close bands (F_2, F_3, and F_4). *M. peregrinum* also produced four bands, in this case occurring in two, well separated pairs.

When the fast fraction S_4 of *M. fortuitum* was eluted from agarose slabs and subsequently run on polyacrylamide gels, a single band corresponding to F_2 was obtained.[284] The slower agarose band (S_3) produced three bands (F_2, F_3, and F_4) on polyacrylamide gels. These observations are significant because the presence of F_2 both in S_3 and in S_4 clearly implies a close relationship between the latter two bands. Although there are many other examples of multiple arylsulfatases in microorganisms, this is one of the few cases where a relationship between isoenzymes has been demonstrated.

The overall similarity of zymograms for arylsulfatases of *M. fortuitum* and *M. peregrinum* (Figure 8) supports an earlier suggestion[286] of a close relationship between these species. Moreover, there is some degree of immunological cross-reactivity between their arylsulfatases.[284] In contrast, the *M. chelonei* enzymes share no immunological identity with *M. fortuitum* and *M. peregrinum* and also produce distinctly different zymograms.[284]

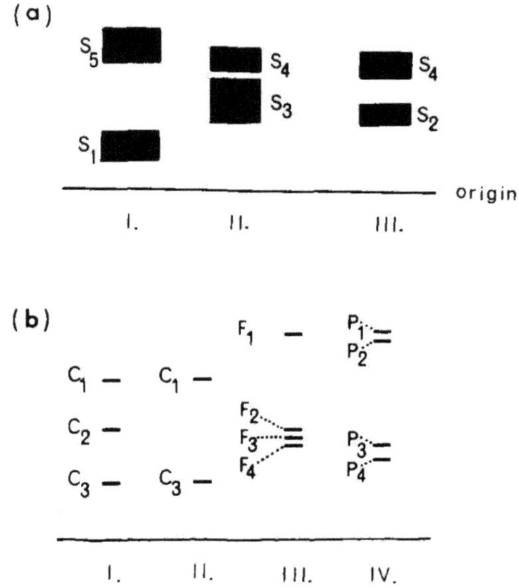

FIGURE 8. Arylsulfatase isoenzymes in some *Mycobacterium* species. (a), Agarose gel zymograms of extracts of *M. chelonei* (I), *M. fortuitum* (II), *M. peregrinum* (III); (b), polyacrylamide gel zymograms of extracts of *M. chelonei* subsp. *chelonei* (I), *M. chelonei* subsp. *chelonei* or *abscessus* (II), *M. fortuitum* (III), *M. peregrinum* (IV). 6-Benzoyl-2-naphthyl sulfate was used to locate arylsulfatase activity.

Finally, arylsulfatases in extracts of *M. fortuitum* collectively exhibit maximum activity at pH 7.[287] However, it should be emphasized that this value was obtained using 4-methylumbelliferyl sulfate and the pattern of multiple enzymes for this substrate might not necessarily correspond with those shown in Figure 8 for 6-benzoyl 2-naphthyl sulfate. Furthermore, if there is a multiplicity of enzymes acting on 4-methylumbelliferyl sulfate, the observed pH optimum only applies to crude extracts, and not necessarily to any particular isoenzyme(s).

V. ALGAL ARYLSULFATASE

A. Arylsulfatase of *Chlamydomonas reinhardti*

C. reinhardti is the only alga that has been investigated for its ability to synthesize arylsulfatase. As in several other instances mentioned in this chapter, the enzyme was purified and characterized[288] as part of a broader study to examine derepressibility of arylsulfatase during the life cycle of this alga.

1. Purification

Cells were derepressed by sulfur deprivation,[289] harvested and acetone-dried. The acetone powder was extracted with phosphate buffer and the extracts clarified by filtration through a glass-fiber filter. The arylsulfatase was purified from the extract by a sequence of ammonium sulfate precipitation (0 to 60%), Sephadex® G-200, DEAE-Sephadex®, and CM-Sephadex® chromatography to give a 21% recovery with 85-fold purification. The protein in the final material was homogeneous according to polyacrylamide gel electrophoresis under acid and alkaline conditions, with coincident migration of protein and enzyme activity. The protein also moved as a single symmetrical peak during sedimentation in the ultracentrifuge.

2. Properties

The molecular weight of the purified enzyme was 150,000 (Table 4) and its isoelectric point was 9.0, implying a large content of basic amino acids.[288] It was not particularly stable, losing 25% of its activity in 14 days at 3°C. In the longer term, the inactivation of the pure enzyme could not be arrested either at low temperatures (90% loss in 8 weeks at −20°C) or by the addition of mercaptoethanol or imidazole. The pure enzyme was rapidly inactivated at 60°C at acid or alkaline pH, but it was more stable at pH 6.5 to 7.0. Perhaps not surprisingly, the enzyme was more stable in intact cells than in the pure state at a given pH. Maximum activity was observed at 30°C for the pure enzyme and the activation energy at the optimum pH of 9.8 was 7.9 kcal/mol.

Although imidazole failed to prevent inactivation, its presence in assay mixtures caused an increase in activity over the whole range of pH with the highest (10 to 15 fold) increase under acid and neutral conditions.[288] The presence of imidazole therefore effectively shifted the optimum pH from 9.8 to 9.0. This effect is reminiscent of the activation and shift in optimum pH produced by chloride for nitrocatechol sulfate hydrolysis by the *P. vulgaris* arylsulfatase (Section IV.B.5) although with the algal enzyme the effect occurred with both *p*-nitrophenyl sulfate and nitrocatechol sulfate. In the presence of 10 m*M* imidazole the K_m for *p*-nitrophenyl sulfate (2×10^{-4} *M*) was about three times the value in the absence of imidazole. Although the presence of imidazole decreased the affinity of enzyme for its substrate, the value of V_{max} was increased sufficiently to produce the observed overall activation.

Sulfate, sulfite, phosphate, methionine, cysteine, citrate, and EDTA all had no effect on the enzyme's activity.[288] Borate was a strong competitive inhibitor (K_i = 5×10^{-5} *M*), a property now known to be shared with some of the arylsulfatases in *Aspergillus* species (Table 7). Cyanide was also a strong inhibitor, but this inhibition was noncompetitive. These inhibitors exerted their effects in the absence and in the presence of imidazole.[288] The inhibition pattern and the preference for *p*-nitrophenyl sulfate as substrate were sufficient to classify the enzyme as a Type I.

The properties of the enzyme investigated using intact cells clearly resembled those found with the pure enzyme.[288] This, together with the ease of assay of the activity in intact cells and the inability to increase the activity by rupturing cells or by preparing acetone powders, suggested that the arylsulfatase in *C. reinhardti* was located near the surface of the cell. However, despite its classical Type I behavior and cell-surface location the enzyme does have some features that are atypical of microbial arylsulfatases. It is not inhibited by sulfite, a property shared with only two other known microbial arylsulfatases (*A. aerogenes* and *C. terrigena*, see Tables 7 and 9). It also has a very alkaline pH optimum comparable to that for the arylsulfatase in *S. alcalophilus* (Table 8), a highly alkaline isoelectric point, and is activated by imidazole. There is no doubt that this algal enzyme stands out from other microbial arylsulfatases and it is to be hoped that this will stimulate further work on algal arylsulfatases.

VI. CONCLUDING REMARKS

A. Comments on the Purification of Arylsulfatases

The foregoing sections reveal that although arylsulfatase enzymes are widely distributed in microorganisms only a few have been purified sufficiently to permit detailed characterization. The purification procedures so far devised generally conform to the usual pattern of batch precipitation procedures (using ammonium sulfate or organic solvents), followed by methods offering greater resolution such as ion-exchange and gel-filtration chromatography. Somewhat surprisingly the powerful technique of affinity chromatography has been used successfully in only one instance, for the arylsulfatase of *K. aerogenes* (Section IV.B.2.a). The affinity media used contained a β-1,3-glucan matrix to which was attached either tyrosine or a specific antibody for the arylsulfatase of *K. aerogenes* W70. Of course, the latter approach suffers from the disadvantage that it requires a supply of pure enzyme in order to raise antibody in the first instance.

Attempts have been made to purify mammalian arylsulfatases using an aryl sulfate (4-hydroxy 2-nitrophenyl sulfate) as affinity ligand.[290] This ester was a poor substrate for mammalian arylsulfatase B and the affinity medium containing it was useful for the purification of that enzyme. However, arylsulfatases of *A. oryzae, A. metalcaligenes,*[290] and *Pseudomonas* C12B[261] did not bind to the column. Derappe and Dubois[291] reported the synthesis of an analogue of nitrocatechol sulfate that would be suitable for attachment to a Sepharose® support. However, these authors did not report the efficacy of the material for purifying arylsulfatases. van Etten and Waheed[292] adopted the novel approach of linking single subunits of arylsulfatase A to a matrix and using this as an affinity medium for other subunits in solution. However, this method is probably not suitable for microbial arylsulfatases which, for the most part, are monomeric proteins.

It may be possible to avoid the need to synthesize affinity chromatography media by using commercially available affinity media for the purification of microbial arylsulfatases. Cibacron Blue Sepharose® has been used successfully for separating, for example, arylsulfatases A and B.[293,294] Presumably the arylsulfatases interact with arylsulfonic acid residues of the affinity ligand. The purification of arylsulfatase A achieved using SP (sulfopropyl)-Sephadex® probably exploits a similar interaction with a sulfonate moiety.[295] It would seem worthwhile investigating the suitability of these materials for the purification of microbial arylsulfatases.

B. Multiplicity of Arylsulfatases

The use of acetone either in the preparation of acetone-dried cells or in the precipitation of protein from crude cell-extracts requires further comment. Frequently the use of acetone in the early stages of purification is associated with detection of arylsulfatase isoenzymes. For example, *A. oryzae,*[153] *A. nidulans,*[168] *S. typhimurium,*[231] *P. rettgeri,*[296] and *P. aeruginosa*[254] all yield multiple arylsulfatases following treatment with acetone (or in the case of *A. nidulans*, ethanol). Dodgson et al.[297] observed that fresh mammalian liver contained a single arylsulfatase but that acetone-dried extracts had a second form. It is now well known[298,299] that mammalian livers possess three arylsulfatases designated A, B, and C. Arylsulfatases A and B are lysosomal enzymes and enzyme C is microsomal. The effect of acetone-drying may therefore be to increase accessibility of one or more enzymes to the substrate rather than producing an artifact. Although this could be the case with some of the microbial enzymes, it would be unwise to make a sweeping generalization. Thus, arylsulfatases were extracted from acetone-dried cells of *A. aerogenes* and *A. metalcaligenes* with no evidence of heterogeneity. Moreover, equally complex isoenzyme patterns were obtained for arylsulfatases from *P. rettgeri* irrespective of whether the cell extracts were prepared from acetone-dried cells or by sonication,[296] or even by using a French pressure cell.[305] Similarly, in a study of urease isoenzymes, Fishbein[300] found that even in filtered aqueous extracts untreated with acetone, there were a number of isoenzymes separable by gel electrophoresis. Moreover, the patterns varied with the source of Jackbean meal from which extracts were prepared. Repeated crystallization of urease from the extracts by treatment with acetone did, however, result in interconversion of certain forms of the enzyme. Thus, although acetone can influence the observed isoenzymic pattern, the appearance of multiple enzymes following treatment with acetone by no means implies that all isoenzyme forms arise as artifacts. Indeed, the occurrence of multiple forms of arylsulfatases in mammalian systems is common, particularly for arylsulfatase B of ox liver,[301] ox brain,[302] and corn meal extracts,[303] even when no organic solvents were employed. The differences among these isoenzymes appeared to lie predominantly in their ionic charges rather than molecular shape or size.[301,302] Aggregation of the ox liver enzyme producing further subforms occurred during electrophoresis at low ionic strength and it was suggested that this could be a common problem during electrophoretic procedures. Possibly the addition of acetone to solutions of arylsulfatase could alter the dielectric properties of water sufficiently to produce a similar aggregation effect. However, little real progress is likely to be forthcoming with these problems until various isoenzymes are isolated in sufficient quantities to allow structural analyses and comparisons to be made.

The structural relationships among microbial arylsulfatase isoenzymes for any particular organism are indeed unclear, but there are indications that such relationships do exist. For example,

1. Arylsulfatase II of *A. oryzae* exists in monomeric and dimeric forms both of which are active and may be related to isoenzymes I and III of the same fungus (Section III.B.2).
2. The two arylsulfatases of *A. nidulans* that differ in surface charge and response to some inhibitors are similar in many other respects and, moreover, they are also produced sequentially suggesting a possible posttranslational modification of the first isoenzyme to produce the second (Section III.B.5).
3. When extracts of *M. fortuitum* are subjected to agarose electrophoresis, two bands appear that in turn reveal fine structures with a common component when reelectrophoresed on polyacrylamide gels (Section IV.F).
4. The arylsulfatase of *A. aerogenes* loses activity in the cold and extra protein peaks appeared when stored solutions were subjected to ion-exchange chromatography (Section IV.B.1).
5. The arylsulfatase of *S. typhimurium* which was homogeneous on polyacrylamide gels, separates into five isoenzyme bands during isoelectric focusing (Section IV.B.3).

With the exception of arylsulfatase II of *A. oryzae* and possibly the arylsulfatase of *N. crassa* mycelia, no other known microbial arylsulfatase has a subunit structure. The enzymes tend to be quite small (mostly under 100,000) particularly those of bacterial origin which all have a molecular weight around 50,000 (Table 4). The possibility that isoenzymes are generated by combining subunits in various numbers is therefore limited, particularly for example in *A. nidulans* where isoenzymes have the same molecular weight under nondenaturing conditions. Mammalian arylsulfatase A isoenzymes have been shown[232] to be due to variation in sialic acid content, but this was not a distinguishing feature for the five isoenzymes of *S. typhimurium*.[233] Whether variation in carbohydrate content accounts for arylsulfatase multiplicity in other microorganisms remains to be seen.

Besides structural considerations, it is also important to consider the implications of arylsulfatase multiplicity in terms of the role of the multiple enzymes in the cell. At the beginning of this chapter emphasis was placed on the abundance of natural arylsulfate esters returning to soil and natural waters, and the requirement of microorganisms for arylsulfatases to enable them to utilize such compounds as sources of carbon and/or sulfur. A possible reason for the existence of multiple arylsulfatases may therefore be to provide the organism with enzymic equipment of wide specificity to permit the utilization of a range of structurally different aryl sulfate esters. This role which is consistent with the association of some microbial arylsulfatases with the cell wall (Volume II, Chapter 4), would seem appropriate, for example, for the inducible arylsulfatase and tyrosine sulfate sulfatase of *C. terrigena.* A second possibility is that not all isoenzymes in a given organism are concerned in vivo with hydrolysis of aryl sulfate esters. Precedent for this can be found again in the mammalian liver arylsulfatases in which the microsomal arylsulfatase C is a true aryl (estrone) sulfatase, but the lysosomal A and B enzymes are concerned in vivo with the hydrolysis of galactose 3-sulfate residues in cerebroside sulfate, and galactosamine 4-sulfate and glucosamine 4,6-disulfate residues, respectively.[143,298,299] This suggestion is particularly relevant for fungal arylsulfatases that appear to be located in different compartments in the cell (see, for example, Section III.C.) including lysosomal granules. Further discussion of the roles of arylsulfatases can be found in Volume II, Chapter 4.

C. The Use of Nitrocatechol Sulfate as Substrate

An interesting catalytic feature of arylsulfatases is their relative activities toward *p*-nitrophenyl sulfate and nitrocatechol sulfate. Chuaqui and Lynn[304] have recently collected relative V_{max} values for a number of arylsulfatases from microbial, molluscan, and mammalian sources and have shown that enzymes for which nitrocatechol sulfate is the better substrate have the more acidic pH optima. In order to rationalize this observation, they supposed that the phenolic form of nitrocatechol sulfate with an unionized OH group served as a substrate but that the phenolate anion did not. Arylsulfatases operating under acidic conditions thus have a higher proportion of the substrate form available to them than those operating under more alkaline conditions. The acidic arylsulfatases therefore show relatively high activity toward nitrocatechol sulfate. Arylsulfatases from several sources including *P. vulgaris, A. oryzae, A. aerogenes, P. rettgeri,* and *A. metalcaligenes* were shown to conform to this pattern. The phenomenon is related to the susceptibility of arylsulfatases to inhibition by inorganic sulfate. In Section II.B.4. of Volume II, Chapter 2, inhibition by sulfate is considered in relation to the mechanism of hydrolysis of aryl sulfate esters. It is therefore concluded on the basis of the pH-dependence of the inhibition that the true inhibitor is probably HSO_4^- and that inhibition by added sulfate only becomes apparent when the pH of the assay mixture is low enough to convert appreciable amounts of SO_4^{2-} to HSO_4^-. Inhibition by sulfate would only be apparent for those enzymes that were most active under, and therefore assayed under, acid conditions. In short, arylsulfatases with acidic pH optima would be expected to be inhibited by sulfate and to show high activity toward nitrocatechol sulfate. Interestingly, these are two of the criteria used to distinguish Type 2 arylsulfatases from the Type 1 enzymes. Because they both seem to be linked to the pH characteristics of arylsulfatase activity, it may be that a revision of the criteria as advocated by Nicholls and Roy[143] could usefully be made along these lines. Finally, with regard to the preceding section, the mammalian Type 2 arylsulfatases A and B with high activity toward nitrocatechol sulfate and acid pH optima are lysosomal enzymes acting in vivo on nonaryl sulfate esters. It may be that microbial enzymes with these characteristics are those for which a search should be instigated for alternative, nonaryl sulfate ester substrates.

REFERENCES

1. **Roy, A. B.,** Sulphatases, lysosomes and disease, *Aust. J. Exp. Biol. Med. Sci.,* 54, 111, 1976.
2. **Roy, A. B.,** Sulphatase deficiencies, in *The Cultured Cell and Inherited Metabolic Disease,* Harkness, R. A. and Cockburn, F., Eds., MTP Press, Lancaster, England, 1977, 120.

3. **Roy, A. B.,** Sulphatase A — an arylsulphatase and a glycosulphatase, in *Sulfur in Biology,* CIBA Foundation Symposium, No. 72, Excerpta Medica, Amsterdam, 1980, 177.
4. **Harada, T. and Murooka, Y.,** Participation of tyramine oxidase in multiple control of bacterial arylsulfatase synthesis, *Mem. Inst. Sci. Ind. Res., Osaka Univ.,* 37, 45, 1980.
5. **Dodgson, K. S. and Rose, F. A.,** Sulphoconjugation and sulphohydrolysis, in *Metabolic Conjugation and Metabolic Hydrolysis,* Vol.1, Fishman, W. H., Ed., Academic Press, New York, 1970, 239.
6. **Dodgson, K. S.,** Conjugation with sulphate, in *Drug Metabolism; from Microbe to Man,* Parke, D. V. and Smith, R. L., Eds., Taylor and Francis, London, 1977, 91.
7. **Powell, G. M. and Olavesen, A. H.,** The fate of sulfate esters in vivo, in *Sulfation of Drugs and Related Compounds,* Mulder, G. J., Ed., CRC Press, Boca Raton, Fl., 1981, chap. 7.
8. **Baumann, E.,** Über Sulfosauren im Harn, *Ber. Deut. Chem. Ges.,* 9, 54, 1876.
9. **Boström, H., Gustafsson, B. E., and Wengle, B.,** Studies on ester sulphates. XVIII. Ester sulphate formation in the germ-free rat, *Proc. Soc. Exp. Biol. Med.,* 114, 742, 1963.
10. **Smith, A. A.,** Origin of urinary pyrocatechol, *Nature (London),* 190, 167, 1961.
11. **Curtis, C. G., Hearse, D., and Powell, G. M.,** Renal excretion of some arylsulphate esters in the rat, *Xenobiotica,* 4, 595, 1974.
12. **Curzon, G. and Pratt, R. T. C.,** Origin of urinary resorcinol sulphate, *Nature (London),* 204, 383, 1964.
13. **Allsop, T. F.,** Comparative Detoxication of Phenols by Some Marine Invertebrates, M.Sc. thesis, University of Wellington, Wellington, New Zealand, 1965.
14. **Corner, E. D. S., Kilvington, C. C., and O'Hara, S. C. M.,** Qualitative studies on the metabolism of naphthalene in *Maia squinada* (Herbst), *J. Mar. Biol. Assoc. U.K.,* 53, 833, 1973.
15. **Boylan, D. B. and Tripp, B. W.,** Determination of hydrocarbons in sea-water extracts of crude oil and crude oil fractions, *Nature (London),* 230, 44, 1971.
16. **Corner, E. D. S. and Young, L.,** Biochemical studies of toxic agents. VII. The metabolism of naphthalene in animals of different species, *Biochem. J.,* 58, 647, 1954.
17. **Terriere, L. C., Boose, R. B., and Roubal, W. T.,** The metabolism of naphthalene and 1-naphthol by houseflies and rats, *Biochem. J.,* 79, 620, 1961.
18. **Smith, J. N. and Turbert, H. B.,** Comparative detoxication. II. Conjugations of 1-naphthol and some other phenols in houseflies and locusts, *Biochem. J.,* 92, 127, 1964.
19. **Sato, T., Suzuki, T., Fukuyama, T., and Yoshikawa, H.,** Studies on conjugation of ^{35}S-sulfate with phenolic compounds. IV. Metabolism of *o*-cresol, *m*-cresol, salicylaldehyde, salicylic acid, toluene, benzoic acid, and related substances in rat liver, *J. Biochem. (Tokyo),* 43, 421, 1956.
20. **Berenbom, M. and Young, L.,** Biochemical studies of toxic agents. III. The isolation of 1- and 2-naphthyl sulfuric acid and 1- and 2-naphthyl glucuronide from the urine of rats dosed with 1- and 2-naphthols, *Biochem. J.,* 49, 165, 1951.
21. **Kotin, P., Falk, H. L., and Busser, R.,** Distribution, retention and elimination of ^{14}C-3,4-benzpyrene after administration to mice and rats, *J. Natl. Cancer Inst.,* 23, 541, 1959.
22. **Bettelheim, F. R.,** Tyrosine *O*-sulfate in a peptide from fibrinogen, *J. Am. Chem. Soc.,* 76, 2838, 1954.
23. **Doolittle, R. F. and Blombäck, B.,** Amino acid sequence investigations of fibrinopeptides from various mammals: evolutionary implications, *Nature (London),* 202, 147, 1964.
24. **Gregory, H., Hardy, P. M., Jones, D. S., Kenner, G. W., and Sheppard, R. C.,** The antral hormone gastrin, *Nature (London),* 204, 931, 1964.
25. **Mutt, V. and Jorpes, J. E.,** Structure of porcine cholecystokinin-pancreozymin. I. Cleavage with thrombin and with trypsin, *Eur. J. Biochem.,* 6, 156, 1968.
26. **Anastasi, A., Bertaccini, G., and Erspamer, V.,** Pharmacological data on phyllokinin (bradykinyl-isoleucyl tyrosine *O*-sulfate) and bradykinylisoleucyl tyrosine, *Br. J. Pharmacol. Chemother.,* 27, 479, 1966.
27. **Anastasi, A., Erspamer, V., and Endean, R.,** Isolation and amino acid sequence of caerulein, the active decapeptide of the skin of *Hyla caerulea, Arch. Biochem. Biophys.,* 125, 57, 1968.
28. **Tallan, H. H., Bella, S. T., Stein, W. H., and Moore, S.,** Tyrosine *O*-sulfate as a constituent of normal human urine, *J. Biol. Chem.,* 217, 703, 1955.
29. **Grimes, A. J.,** Synthesis of ^{35}S-labelled arylsulphates by intact animals and by tissue preparations with particular reference to L-tyrosine *O*-sulfate, *Biochem. J.,* 73, 723, 1959.
30. **John, R. A., Rose, F. A., Wusteman, F. S., and Dodgson, K. S.,** The detection and determination of L-tyrosine *O*-sulphate in rabbit and other mammalian urine, *Biochem. J.,* 100, 278, 1966.
31. **Hext, P. M., Thomas, S., Rose, F. A., and Dodgson, K. S.,** Determination and significance of tyrosine *O*-sulfate and its deaminated metabolites in normal human and mouse urine, *Biochem. J.,* 134, 629, 1973.
32. **Kittredge, J. S., Simonsen, D. G., Roberts, E., and Jelinek, B.,** Free amino acids of marine invertebrates, in *Amino Acid Pools,* Holden, J. T., Ed., Elsevier, London, 1962, 176.

33. **Roche, J., Michel, R., Closon, J., and Michel, O.**, Sur la sulfoconjugation hépatique de la 3,5,3′-triiodo-L-thyronine et la présence d'un ester sulfurique de cette hormone dans la bile et la plasma, *Biochim. Biophys. Acta*, 33, 461, 1959.
34. **Cohn, G. L.**, Biosynthesis of triiodothyronine sulphate by beef thyroid in vitro, *Nature (London)*, 208, 80, 1965.
35. **Boström, H. and Wengle, B.**, Studies on ester sulphate. XIX. On sulphate conjugation in adult human liver extracts, *Acta Soc. Med. Ups.*, 69, 41, 1964.
36. **Goodall, McC. and Alton, H.**, Metabolism of 3,4-dihydroxyphenylalanine (L-DOPA) in human subjects, *Biochem. Pharmacol.*, 21, 2401, 1972.
37. **Kuchel, O., Buu, N. T., and Unger, T.**, Free and conjugated dopamine; physiological and clinical implications, *Adv. Biosci.*, 20, 15, 1979.
38. **Buu, N. T. and Kuchel, O.**, A new method for the hydrolysis of conjugated catecholamines, *J. Lab. Clin. Med.*, 90, 680, 1977.
39. **Jenner, W. N. and Rose, F. A.**, Dopamine 3-*O*-sulfate, an end product of L-dopa metabolism in Parkinson patients, *Nature (London)*, 252, 237, 1974.
40. **Bronaugh, R. L., Hattox, S. E., Hoehn, M. M., Murphy, R. C., and Rutledge, C. O.**, Separation and identification of dopamine 3-*O*-sulfate and 4-*O*-sulfate in urine in Parkinsonian patients, *J. Pharmacol. Exp. Ther.*, 195, 441, 1975.
41. **Gey, K. F. and Pletscher, A.**, Distribution and metabolism of DL-3,4-dihydroxy (2-^{14}C) phenylalanine in rat tissues, *Biochem. J.*, 92, 300, 1964.
42. **Bodnaryk, R. P. and Brunet, P. C. J.**, 3-*O*-Hydrosulphato-4-hydroxyphenethylamine (dopamine 3-*O*-sulphate), a metabolite involved in the sclerotization of insect cuticle, *Biochem. J.*, 138, 463, 1974.
43. **Kennedy, M. B.**, Products of biogenic amine metabolism in the lobster: sulfate conjugates, *J. Neurochem.*, 30, 315, 1978.
44. **Goodall, McC. and Alton, H.**, Metabolism of 3-hydroxytyramine (dopamine) in human subjets, *Biochem. Pharmacol.*, 17, 905, 1968.
45. **Schanberg, S. M., Schildkraut, J. J., Breese, G. R., and Kopin, I. J.**, Metabolism of normetanephrine-^{3}H in rat brain — identification of conjugated 3-methoxy-4-hydroxyphenyl-glycol as the major metabolite, *Biochem. Pharmacol.*, 17, 247, 1968.
46. **Schanberg, S. M., Breese, G. R., Schildkraut, J. J., Gordon, E. K., and Kopin, I. J.**, 3-Methoxy 4-hydroxyphenylglycol sulfate in brain and cerebrospinal fluid, *Biochem. Pharmacol.*, 17, 2006, 1968.
47. **Sugden, R. F. and Eccleston, D.**, Glycol sulphate ester formation from (^{14}C)-noradrenaline in braine and the influence of COMT inhibitors, *J. Neurochem.*, 18, 2461, 1971.
48. **Meek, J. L. and Neff, N. H.**, The rate of formation of 3-methoxy 4-hydroxyphenylethylene glycol sulfate in brain as an estimate of the rate of formation of norepinephrine, *J. Pharmacol. Exp. Ther.*, 184, 570, 1973.
49. **Lopachin, R. M. and Reigle, T. G.**, The effects of several narcotic analgesics on brain levels of 3-methoxy-4-hydroxyphenylethylene glycol sulfate in the rat, *J. Pharmacol. Exp. Ther.*, 207, 151, 1978.
50. **Scudi, J. V., Buhs, R. P., and Hood, D. B.**, The metabolism of vitamin B_6, *J. Biol. Chem.*, 142, 323, 1942.
51. **Sata, T., Yamada, M., Suzuki, T., Fukuyama, T., and Yoshikawa, H.**, Studies on conjugation of ^{35}S-sulfate with phenolic compounds. V. Conjugation of sulfate with pyridoxine and pyridoxal in rat liver, *J. Biochem. (Tokyo)*, 46, 79, 1959.
52. **Baker, J. T.**, Tyrian purple: an ancient dye, a modern problem, *Endeavour*, 33, 11, 1974.
53. **Bouchilloux, J. and Roche, J.**, Contribution to the biochemical study of *Murex* purple, *Bull. Soc. Chim. Biol.*, 37, 37, 1955.
54. **Baker, J. T. and Sutherland, M. D.**, Pigments of marine animals. VIII. Precursors of 6,6′-dibromoindigotin (Tyrian Purple) from the mollusk *Dicathais orbita* Gmelin, *Tet. Lett.*, 1, 43, 1968.
55. **Baker, J. T. and Duke, C. C.**, Isolation of choline and choline ester salts of tyrindoxyl sulfate from the marine mollusks *Dicathais orbita* and *Mancinella keineri*, *Tet. Lett.*, 15, 1233, 1976.
56. **Fouquet, H. and Bielig, H.-J.**, Biological precursors and the genesis of Tyrian purple, *J. Angew. Chemie*, 10, 816, 1971.
57. **Davis, V. E., Huff, J. A., and Brown, H.**, Isolation of serotonin-*O*-sulfate from human urine, *Clin. Chim. Acta*, 13, 380, 1966.
58. **Chadwick, B. T. and Wilkinson, J. H.**, Some aspects of the metabolism of 5-hydroxytryptamine, *Biochem. J.*, 76, 102, 1960.
59. **Heacock, R. A. and Mahon, M. E.**, Conjugated hydroxyskatoles in human urine, *Can. J. Biochem.*, 42, 813, 1964.
60. **Dalgliesh, C. E.**, Excretion of conjugated 2-amino-3-hydroxyacetophenone by man and its significance in tryptophan metabolism, *Biochem. J.*, 61, 334, 1955.
61. **Wieland, H. and Vocke, F.**, Über die Giftstoffe de japanischen Kröte, *J. Liebig's Ann. Chem.*, 481, 215, 1930.

62. **Jensen, H.**, Chemical studies on toad poisons. VII. *Bufo arenum, Bufo regularis* and *Xenopus laevis, J. Am. Chem. Soc.*, 57, 1765, 1935.
63. **Erspamer, V.**, Isolation of bufoviridine from the skin of *Bufo viridis* and its identification as dihydrobufothionine, *Biochem. Pharmacol.*, 2, 270, 1959.
64. **Erspamer, V., Vitali, T., Roseghini, M., and Cei, J. M.**, 5-Methoxy-and 5-hydroxy-indoles in the skin of *Bufo alvarius, Biochem. Pharmacol.*, 16, 1149, 1967.
65. **Butenandt, A., Biekert, E., Kübler, H., and Linzen, B.**, Über Ommochrome. XX. Zur Verbreitung der Ommatin in Tierreich Neue Methoden zu ihrer Identifizierung und quantitativen Bestimmung, *Hoppe-Seyler's Z. Physiol. Chem.*, 319, 238, 1960.
66. **Butenandt, A., Biekert, E., Koga, N., and Traub, P.**, Über Ommochrome. XXI. Konstitution und Syntheses des Ommatins D, *Hoppe-Seyler's Z. Physiol. Chem.*, 321, 258, 1960.
67. **McKenna, J., Menini, E., and Normyberski, J. K.**, Oestrone sulphate in human pregnancy urine, *Biochem. J.*, 79, 11P, 1961.
68. **Schacter, B. and Marrian, G. F.**, The isolation of estrone sulfate from the urine of pregnant mares, *J. Biol. Chem.*, 126, 663, 1938.
69. **Purdy, R. H., Engel, L. L., and Oncley, J. L.**, The characterization of estrone sulfate from human plasma, *J. Biol. Chem.*, 236, 1043, 1961.
70. **Creange, J. E. and Szego, C. M.**, Sulphation as a metabolic pathway for oestradiol in the sea urchin *Strongylocentrotus franciscanus, Biochem. J.*, 102, 898, 1967.
71. **Troen, P., Nilsson, B., Wiqvist, N., and Diczfalusy, E.**, Pattern of oestriol conjugates in human cord blood, amniotic fluid and urine of newborns, *Acta Endocrinol.*, 38, 361, 1961.
72. **Smith, E. R. and Kellie, A. E.**, Oestrogen conjugates of human late pregnancy urine, *Biochem. J.*, 104, 83, 1967.
73. **Touchstone, J. C., Greene, J. W., McElroy, R. C., and Murawee, T.**, Blood estriol conjugation during human pregnancy, *Biochemistry*, 2, 653, 1963.
74. **Menini, E. and Diczfalusy, E.**, Isolation and identification of sodium estriol-3 sulfate in human meconium, *Endocrinology*, 68, 492, 1961.
75. **Sneddon, A. and Marrian, G. F.**, Sulphation of oestrone *in vitro* by bovine adrenal tissue, *Biochem. J.*, 86, 385, 1963.
76. **Adams, J. B.**, Enzymic synthesis of steroid sulphates. I. Sulphation of steroids by human adrenal extracts, *Biochim. Biophys. Acta*, 82, 572, 1964.
77. **Yang, R. S. H. and Wilkinson, C. F.**, Enzymic sulphation of *p*-nitrophenol and steroids by larval gut tissues of the southern armyworm *(Prodenia eridania* Cramer), *Biochem. J.*, 130, 487, 1972.
78. **Kawaguchi, R. and Kim, K. W.**, The constituents of *Persicaria hydropiper, J. Pharm. Soc. Jpn.*, 57, 108, 1937.
79. **Harborne, J. B.**, Flavonoid sulfates: a new class of sulphur compounds in higher plants, *Phytochemistry*, 14, 1147, 1975.
80. **Harborne, J. B.**, Flavonoid sulfates: a new class of natural product of ecological significance in plants, *Prog. Phytochem.*, 4, 189, 1977.
81. **Harborne, J. B. and Williams, C. A.**, Occurrence of sulfated flavones and caffeic acid esters in members of the Fluviales, *Biochem. Syst. Ecol.*, 4, 37, 1976.
82. **Harborne, J. B. and Williams, C. A.**, Flavonoid patterns in leaves of *Gramineae, Biochem. Syst. Ecol.*, 4, 267, 1976.
83. **Harborne, J. B.**, Variation in and functional significance of phenolic conjugation in plants, in *Recent Advances in Phytochemistry*, Vol. 12, Swain, T., Harborne, J. B., and van Sumere, C. F., Eds., Plenum Press, New York, 1978, 457.
84. **Harborne, J. B.**, Flavonoid bisulfates and their co-occurrences with ellagic acid in the *Bixaceae, Frankeniaceae* and related families, *Phytochemistry*, 14, 1331, 1975.
85. **Saleh, N. A. M., Bohm, B. A., and Ornduff, B.**, Flavonoids of *Lasthenia conjugens* and *Lasthenia fremontii, Phytochemistry*, 10, 611, 1971.
86. **Cabrera, J. L. and Juliani, H. R.**, Two new quercetin sulphates from leaves of *Flaveria bidentis, Phytochemistry*, 18, 510, 1979.
87. **Cabrera, J. L. and Juliani, H. R.**, Isorhamnetin-3,7-disulphate from *Flaveria bidentis, Phytochemistry*, 16, 400, 1977.
88. **Nawwar, M. A. M., Ishak, M. S., El Sherbieny, A. E. A., and Meshaal, S. A.**, Flavonoids of *Reaumuria mucronata* and *Thymelaea hirsuta, Phytochemistry*, 16, 1319, 1977.
89. **Saleh, N. A. M., El Sissi, H. I., and Nawwar, M. A. M.**, A rhamnetin glucuronide trisulphate from the leaves of *Tamarix aphylla, Phytochemistry*, 14, 312, 1975.
90. **Cabrera, J. L. and Juliani, H. R.**, Quercetin-3-acetyl-7,3′,4′-trisulphate from *Flaveria bidentis, Lloydia*, 39, 253, 1976.
91. **Pereyra de Santiago, O. J. and Juliani, H. R.**, Isolation of quercetin-3,7,3′,4′-tetrasulfate from *Flaveria bidentis* L. Otto Kuntze, *Experientia*, 28, 380, 1972.

92. **Williams, C. A., Harborne, J. B., and Crosby, T. S.,** Tricetin, diosmetin and luteolin sulphates in leaves of *Lachenalia unifolia, Phytochemistry,* 15, 349, 1976.
93. **Augier, J. and Henry, M.-H.,** Au sujet du brome dans les Rhodophycées, *Bull. Soc. Botan. Fr.,* 97, 29, 1950.
94. **Augier, J. and Mastagli, P.,** Sur un composé phénolique bromé extrait de l'algue rouge, *Halopitys incurvus, C. R. Acad. Sci. Paris Ser. C,* 242, 190, 1956.
95. **Hodgkin, J. H., Craigie, J. S., and McInnes, A. G.,** The occurrence of 2,3-dibromobenzyl alcohol-4,5-disulfate, dipotassium salt, in *Polysiphonia lanosa, Can. J. Chem.,* 44, 74, 1966.
96. **Kurata, K. and Amiya, T.,** Chemical studies on constituents of marine algae. II. Constituents of the red marine alga, *Rhodomela subfusca, Bull. Jpn. Soc. Sci. Fish.,* 41, 657, 1975.
97. **Glombitza, K. W. and Stoffelen, H.,** Antibiotics from alga. VI. 2,3-Dibromo-4,5-dihydroxybenzl alcohol-4,5-bis (hydrogen sulfate) dipotassium salt from *Rhodomelaceae, Planta Med.,* 22, 391, 1972.
98. **Kurata, K., Amiya, T., and Yabe, K.,** Constituents of a red marine alga, *Odonthalia corymbifera, Bull. Jpn. Soc. Sci. Fish.,* 39, 961, 1973.
99. **Glombitza, K. W., Stoffelen, H., Murawski, U., Bielaczek, J., and Egge, H.,** Antibiotics from algae. IX. Bromophenols from *Rhodomelaceae, Planta Med.,* 25, 105, 1974.
100. **Weinstein, B., Rold, T. L., Harrell, C. E., Jr., Burns, M. W., and Waaland, J. R.,** Re-examination of the bromophenols in the red alga, *Rhodomela larix, Phytochemistry,* 14, 2667, 1975.
101. **Pedersén, M., Saenger, P., and Fries, L.,** Simple brominated phenols in red alga, *Phytochemistry,* 13, 2273, 1974.
102. **Fries, L.,** Growth stimulating effects of the bromophenol, lanosol, on red algae in axenic culture, *Experientia,* 29, 1436, 1973.
103. **Mautner, H. G., Gardner, G. M., and Pratt, R.,** Antibiotic activity of seaweed extracts. II. *Rhodomela larix. J. Am. Pharm. Assoc. Sci. Ed.,* 42, 294, 1953.
104. **Péguy, M.,** A propos de l'activité antibiotique présumée d'une solution aqueuse du corps phenolique bromé et sulfoné extrait du *Polysiphonia lanosa* (Linné) Tandy (Floridée *Rhodomélacée) C.R. Acad. Sci.,* 252, 2134, 1961.
105. **McLachlan, J. and Craigie, J. S.,** Antialgal activity of some simple phenols, *J. Phycol.,* 2, 133, 1966.
106. **Stapley, E. O., Jackson, M., Hernandez, S., Zimmerman, S. B., Currie, S. A., Mochales, S., Mata, J. M., Woodruff, H. B., and Hendlin, D.,** Cephamycins, a new family of β-lactam antibiotics I. Production by *Actinomycetes,* including *Streptomyces lactamdurans* sp. n., *Antimicrob. Agric. Chemother.,* 2, 122, 1972.
107. **Albers-Schonberg, G., Arison, B. H., and Smith, J. L.,** New β-lactam antibiotics: structure determination of cephamycins A and B, *Tet. Lett.,* 29, 2911, 1972.
108. **Jensen, A. and Ragan, M. A.,** 1,2,3,5-Tetrahydroxybenzene 2,5-disufate ester; the "phenolic precursor" in Gelbstoff-forming exudates from the marine brown alga, *Ascophyllum nodosum* (L.) Lejol, *Tet. Lett.,* 9, 847, 1978.
109. **Ragan, M. A. and Jensen, A.,** Widespread distribution of sulfated polyphenols in brown algae, *Phytochemistry,* 18, 261, 1979.
110. **Ragan, M. A.,** Phenol sulfate esters: ultraviolet, infrared, ^{1}H and ^{13}C nuclear magnetic resonance spectroscopic investigation, *Can. J. Chem.,* 56, 2681, 1978.
111. **Kent, R. A., Smith, I. R., and Sutherland, M. D.,** Pigments of marine animals. X. Substituted naphthopyrones from the crinoid *Comantheria perplexa, Aust. J. Chem.,* 23, 2325, 1970.
112. **Smith, I. R. and Sutherland, M. D.,** Pigments of marine animals. XI. Angular naphthopyrones from the crinoid *Comanthus parvicirrus timorensis, Aust. J. Chem.,* 24, 1487, 1971.
113. **Ruelius, H. W. and Gauhe, A.,** Über einen Oxynaphthochinon-Schwefelsäure-Ester aus Fusarien, *Liebigs Ann.,* 570, 121, 1950.
114. **Cormier, M. J.,** Applications of *Renilla* bioluminescence: An introduction, *Meth. Enzymol.,* 57, 237, 1978.
115. **Cormier, M. J., Hori, K., Karkhanis, Y. D., Anderson, J. M., Wampler, J. E., Morin, J. G., and Hastings, J. W.,** Evidence for similar biochemical requirements for bioluminescence among the coelenterates, *J. Cell Physiol.,* 81, 291, 1973.
116. **Cormier, M. J., Hori, K., and Anderson, J. M.,** Bioluminescence in coelenterates, *Biochim. Biophys. Acta,* 346, 137, 1974.
117. **Derrien, M.,** Meeting of the Société Chimique de France, Montpellier, 16 December 1910, *Bull. Soc. Chim.,* 9, 110, 1911.
118. **Neuberg, C. and Kurono, K.,** Sulfatases I. The enzymic fission of benzenesulfonic acid (Phenolsulphatase of *Aspergillus oryzae), Biochem. Z.,* 140, 295, 1923.
119. **Lichtenstein, N.,** The proteolytic enzymes of Takadiastase, in *Studies in Biochemistry,* Kisch, B., Ed., Brooklyn Medical Press, New York, 1947, 182.
120. **Neuberg, C. and Lindhardt, K.,** Über Sulfatase. II., *Biochem. Z.,* 142, 191, 1923.
121. **Neuberg, C. and Wagner, J.,** Sulfatases VII. The splitting of ethereal sulfuric acids of different series by sulfatase and the mechanism of the action of sulfatase, *Biochem. Z.,* 161, 492, 1925.

122. **Noguchi, J.**, Über Sulfatase. IV. Mitteilung. Über die Enzymatische spaltung der β-Naphthol-Schwefelsäure, *Biochem. Z.*, 144, 138, 1924.
123. **Fromageot, C.**, Die Stereochemische spezifität der Sulfatase, *Biochem. Z.*, 208, 482, 1929.
124. **Weinmann, F.**, Über Sulfatase. X., *Biochem. Z.*, 205, 214, 1929.
125. **Neuberg, C. and Simon, E.**, Sulfatase and its substrate, *Ergeb. Physiol.*, 34, 896, 1932.
126. **Fujita, S. and Hosoda, Y.**, Sulfatase, *Chem. Abstr.*, 28, 1060, 1934.
127. **Morimoto, K.**, Über die Phenolsulfatase, *J. Biochem. Tokyo*, 26, 259, 1937.
128. **Abbott, L. D.**, Phenolsulfatase activity, *Arch. Biochem. Biophys.*, 15, 205, 1947.
129. **Dzialoszynski, L. M.**, The sulphatase of Clarase, *Nature (London)*, 160, 464, 1947.
130. **Huggins, C. and Smith, D. R.**, Chromogenic substrates. III. *p*-Nitrophenyl sulfate as a substrate for the assay of phenolsulfatase activity, *J. Biol. Chem.*, 170, 391, 1947.
131. **Cohen, H. and Bates, R. W.**, Hydrolysis of conjugated sulfates of estrogens by commercial enzyme preparation of *Aspergillus oryzae*, *Endocrinology*, 44, 317, 1949.
132. **Dzialoszynski, L. M.**, The sulphatase of "clarase": inhibition, inactivation, and purification, *Nature (London)*, 166, 157, 1950.
133. **Robinson, D., Spencer, B., and Williams, R. T.**, Spectrophotometric determination of phenolsulphatase (arylsulphatase), *Biochem. J.*, 48, 28p, 1951.
134. **Robinson, D., Smith, J. N., Spencer, B., and Williams, R. T.** , Studies in detoxication. XLIII. A study of arylsulphatase activity of Takadiastase towards some phenolic ethereal sulphates, *Biochem. J.*, 51, 202, 1952.
135. **Katzman, P. A., Straw, R. F., Buehler, H. J., and Doisy, E. A.**, Hydrolysis of conjugated estrogens, *Rec. Prog. Horm. Res.*, 9, 45, 1954.
136. **Boyland, E., Manson, D., Sims, P., and Williams, D. C.**, The biochemistry of aromatic amines. The resistance of some *o*-amino arylsulfates to hydrolysis by arylsulfatase, *Biochem. J.*, 62, 68, 1956.
137. **Dodgson, K. S., Rose, F. A., and Tudball, N.**, Studies on sulphatases. XXIII. The enzymic desulphation of tyrosine *O*-sulphate, *Biochem. J.*, 71, 10, 1959.
138. **Ney, K. H. and Ammon, R.**, Die Verbreitung der aryl-und steroid sulfatasen, *Z. Physiol. Chem.*, 315, 145, 1959.
139. **Whitehead, J. E. M., Morrison, A. R., and Young, L.**, Bacterial arylsulfatase, *Biochem. J.*, 51, 585, 1952.
140. **Noguchi, J.**, Sulfatase III. Enzymic fission of aromatic ethereal sulfate compounds in horse, sheep and rabbit urines, *Biochem. Z.*, 143, 186, 1923.
141. **Butenandt, A. and Hofstetter, H.**, Estrone sulfate, a physiological excretory product from follicular hormone, *Z. Physiol. Chem.*, 259, 222, 1939.
142. **Dodgson, K. S. and Spencer, B.**, Assay of sulfatases, *Meth. Biochem. Anal.*, 4, 211, 1957.
143. **Nicholls, R. G. and Roy, A. B.**, Arylsulfatases, in *The Enzymes*, Vol. 5, 3rd ed., Boyer, P., Ed., Academic Press, New York, 1971, chap. 2.
144. **Harada, T.**, A new bacterium in urine which makes lignin red. VIII. Hydrolysis of indican in urine by microorganisms, *J. Agric. Chem. Soc. Jpn.*, 26, 95, 1952.
145. **Harada, T., Kono, K., and Yagi, K.**, The methods of examination of phenolsulphatase in microorganisms and the complexity of the enzymes, *Mem. Inst. Sci. Ind. Res., Osaka*, 11, 193, 1954.
146. **Harada, T., Kono, K., and Yagi, K.**, The methods of determining phenolsulphatase in microorganisms and the complex nature of the enzyme, *J. Agric. Chem. Soc. Jpn.*, 28, 337, 1954.
147. **Harada, T. and Spencer, B.**, The effect of sulphate assimilation on the induction of arylsulphatase synthesis in fungi, *Biochem. J.*, 82, 148, 1962.
148. **Young, R. L.**, Purification of fungal phenolsulphatase, *Fed. Proc. Fed. Am. Soc. Exp. Biol.*, 14, 310, 1955.
149. **Young, R. L.**, Purification of fungal phenolsulphatase, *Proc. Soc. Exp. Biol. Med.*, 99, 530, 1958.
150. **Cherayil, J. D. and van Kley, H.**, Studies on the arylsulfatase of *Aspergillus oryzae*, *Fed. Proc. Fed. Am. Soc. Exp. Biol.*, 20, 235, 1961.
151. **Cherayil, J. D. and van Kley, H.**, Multiple arylsulfatases of *Aspergillus oryzae*, *Fed. Proc. Fed. Am. Soc. Exp. Biol.*, 21, 230, 1962.
152. **Cherayil, J. D. and van Kley, H.**, Arylsulfatase isoenzymes of *Aspergillus oryzae*, *Fed. Proc. Fed. Am. Soc. Exp. Biol.*, 22, 241, 1963.
153. **Cherayil, J. D.**, Studies on arylsulfatases of *Aspergillus oryzae*, Ph.D. thesis, St. Louis University, St. Louis, Mo., 1963.
154. **Drnec, J. F.**, The Arylsulfatases of *Aspergillus oryzae*: Purification and Relationship of Isoenzymes II and III, Ph.D. thesis, St. Louis University, St. Louis, Mo., 1968.
155. **Drnec, J. and van Kley, H.**, The structural relationship of arylsulfatase isoenzymes of *Aspergillus oryzae*, *Fed. Proc. Fed. Am. Soc. Exp. Biol.*, 27, 590, 1968.
156. **Presslitz, J. E.**, Studies on Arylsulfatases, Ph.D. thesis, St. Louis University, St. Louis, Mo., 1968.
157. **Benkovic, S. J., Vergara, E. V., and Hevey, R. C.**, Purification and properties of an arylsulphatase from *Aspergillus oryzae*, *J. Biol. Chem.*, 246, 4926, 1971.

158. **Burns, G. R. J. and Wynn, C. H.**, Studies on the arylsulphatase and phenol sulphotransferase activities of *Aspergillus oryzae, Biochem. J.*, 149, 697, 1975.
159. **Burns, G. R. J., Galanopoulou, E., and Wynn, C. H.**, Kinetic studies on phenol sulphate-phenol sulphotransferase of *Aspergillus oryzae, Biochem. J.*, 167, 223, 1977.
160. **Ohtsuki, K., Sakurai, Y., and Shiota, H.**, Three distinct forms of arylsulfatase produced by *Aspergillus* fungi, *J. Ferment. Technol.*, 55, 476, 1977.
161. **Sakurai, Y., Isobe, K., and Shiota, H.**, Partial purification and some properties of an alkaline arylsulfatase produced by *Aspergillus* fungi, *Agric. Biol. Chem.*, 44, 1, 1980.
162. **Burns, G. R. J. and Wynn, C. H.**, A rapid direct assay for the determination of the separate activities of the three arylsulphatases of *Aspergillus oryzae, Biochem. J.*, 166, 411, 1977.
163. **Benkovic, S. J. and Fedor, J. M.**, An active-site titrant for arylsulfate sulfohydrolase, *J. Am. Chem. Soc.*, 94, 8928, 1972.
164. **Sampson, E. J., Vergara, E. V., Fedor, J. M., Funk, M. O., and Benkovic, S. J.**, On the physical properties and mechanism of action of arylsulfate sulfohydrolase II from *Aspergillus oryzae, Arch. Biochem. Biophys.*, 169, 372, 1975.
165. **Nichol, L. W. and Roy, A. B.**, The sulfatase of ox liver. X. Some observations on the intermolecular bonding in sulfatase A, *Biochemistry*, 5, 1379, 1966.
166. **Rasburn, M. and Wynn, C. H.**, Purification and properties of an isoenzyme of arylsulphatase from *Aspergillus oryzae, Biochim. Biophys. Acta*, 293, 191, 1973.
167. **Nichol, L. W. and Roy, A. B.**, The sulfatase of ox liver. IX. The polymerization of sulfatase A, *Biochemistry*, 4, 386, 1965.
168. **Apte, B. N. and Siddiqi, O.**, Purification and properties of arylsulphatase of *Aspergillus nidulans, Biochim. Biophys. Acta*, 242, 129, 1971.
169. **Spencer, B.**, Studies on sulphatases. XXV. Determination of $BaS^{16}O_3{}^{18}O$ by infrared spectroscopy, *Biochem. J.*, 73, 442, 1959.
170. **Apte, B. N., Bhavsar, P. N., and Siddiqi, O.**, Regulation of arylsulfatase in *Aspergillus nidulans, J. Mol. Biol.*, 86, 637, 1974.
171. **Reiss, J.**, Cytochemical detection of hydrolases in fungus cells. III. Arylsulfatase, *J. Histochem. Cytochem.*, 22, 183, 1974.
172. **Scott, W. A. and Metzenberg, R. L.**, Location of arylsulfatase in conidia and young mycelia of *Neurospora crassa, J. Bacteriol.*, 104, 1254, 1970.
173. **Metzenberg, R. L., Chen, G. S., and Ahlgren, S. K.**, Reversion of arylsulfataseless mutants of *Neurospora, Genetics*, 68, 359, 1971.
174. **Metzenberg, R. L. and Parson, J. W.**, Altered repression of some enzymes of sulfur utilization in a temperature-conditional lethal mutant of *Neurospora, Proc. Natl. Acad. Sci. U.S.A.*, 55, 629, 1966.
175. **Scott, W. A.**, Location of Arylsulfatase in *Neurospora crassa*, Ph.D. thesis, University of Wisconsin, Madison, Wis., 1970.
176. **Schmit, J. C. and Brody, S.**, Biochemical genetics of *Neurospora crassa* conidial germination, *Bacteriol. Rev.*, 40, 1, 1976.
177. **Arnold, W. N.**, Heat inactivation kinetics of yeast β-fructo furanosidase: a polydisperse system, *Biochim. Biophys. Acta*, 178, 347, 1969.
178. **Gascón, S., Neumann, N. P., and Lampen, J. O.**, Comparative study of the properties of the purified internal and external invertase from yeast, *J. Biol. Chem.*, 243, 1573, 1968.
179. **Scott, W. A., Munkres, K. D., and Metzenberg, R. L.**, A particulate fraction from *Neurospora crassa* exhibiting arylsulfatase activity, *Arch. Biochem. Biophys.*, 142, 623, 1971.
180. **Günther, Th., Kattner, W., and Merker, H. J.**, Über das Verhalten und die Lokalisation der sauren Phosphatase von Hefezellen bei Repression und Derepression, *Exp. Cell Res.*, 45, 133, 1967.
181. **Narkates, A. J., Montes, L. F., and Wilborn, W. H.**, Biochemical and cytochemical correlation of acid phosphatase activity with phases of growth in *Candida albicans, J. Histochem. Cytochem.*, 16, 513, 1968.
182. **Pitt, D.**, Histochemical demonstration of certain hydrolytic enzymes within cytoplasmic particles of *Botrytis cinerea* Fr., *J. Gen. Microbiol.*, 52, 67, 1968.
183. **Wilson, C. L., Stiers, D. L., and Smith, G. G.**, Fungal lysosomes or spherosomes, *Phytopathology*, 60, 216, 1970.
184. **Metzenberg, R. L.**, Repair of multiple defects of a regulatory mutant of *Neurospora* by high osmotic pressure and by reversion, *Arch. Biochem. Biophys.*, 125, 532, 1968.
185. **Marzluf, G. A.**, Control of the synthesis, activity, and turnover of enzymes of sulfur metabolism in *Neurospora crassa, Arch. Biochem. Biophys.*, 150, 714, 1972.
186. **McGuire, W. G. and Marzluf, G. A.**, Developmental regulation of choline sulfatase and aryl sulfatase in *Neurospora crassa, Arch. Biochem. Biophys.*, 161, 360, 1974.
187. **Marzluf, G. A. and Metzenberg, R. L.**, Positive control by the *cys*-3 locus in the regulation of sulfur metabolism in *Neurospora, J. Mol. Biol.*, 33, 423, 1968.

188. **Burton, E. G. and Metzenberg, R. L.**, Arylsulfatase in ascospores of *Neurospora crassa, J. Bacteriol.*, 113, 519, 1973.
189. **Burton, E. G. and Metzenberg, R. L.**, Novel mutation causing derepression of several enzymes of sulfur metabolism in *Neurospora crassa, J. Bacteriol.*, 109, 140, 1972.
190. **Armentrout, V. N. and Wilson, C. L.**, Haustorial-host interaction during mycoparasitism of *Mycotypha microspora* by *Piptocephalis virginiana, Phytopathology*, 59, 897, 1969.
191. **Pitt, D. and Combes, C.**, The disruption of lysosome-like particles of *Solanum tuberosum* cells during infections by *Phytophthora erythroseptica, J. Gen. Microbiol.*, 53, 197, 1968.
192. **Dennen, D. W. and Carver, D. D.**, Sulfatase regulation and antibiotic synthesis in *Cephalosporium acremonium, Can. J. Microbiol.*, 15, 175, 1969.
193. **Milazzo, F. H. and Lougheed, G. J.**, An arylsulfatase in the wood-rotting Basidiomycete, *Fomes geotropus, Can. J. Bot.*, 45, 532, 1967.
194. **Reiss, J.**, Particulate intracellular localization of acid phosphatase and aryl sulfatase in the cells of *Rhodotorula rubra, Cytobiologie*, 8, 19, 1973.
195. **Garrison, R. G., Mariat, F., Boyd, K. S., and Tally, J. F.**, Ultrastructure and electron cytochemical studies of *Entomophthora coronata, Ann. Microbiol. (Paris)*, 126B, 149, 1975.
196. **Hopsu-Havu, V. K. and Tunnela, E.**, Production of elastase, urease and sulfatase by *Epidermophyton floccosum* (Harz) Langeron and Milochevitch, *Mykosen*, 20, 91, 1977.
197. **Hoppe-Seyler, G.**, Indigo-forming substances in urine. II. Decomposition of indoxylsulfate in urine by bacteria and the origin of so called "indiguria", *Z. Physiol. Chem.*, 97, 250, 1916.
198. **Harada, T.**, On a new bacterium which colours lignin red in urine, *Mem. Inst. Sci. Ind. Res. Osaka Univ.*, 6, 79, 1948.
199. **Harada, T.**, A new bacterium in urine which makes lignin red. I. Outline of the phenomenon. II. Isolation of the bacterium and its general properties. III. Specificity of the lignin coloration and the probable original coloring substance in urine, *J. Agric. Chem. Soc. Jpn.*, 22, 93, 1948.
200. **Harada, T. and Nishikawa, E.**, On a new bacterium in urine which makes lignin red. IV. Conditions of lignin coloration and the decision of the toluene soluble purple pigments, *J. Agric. Chem. Soc. Jpn.*, 23, 92, 1949.
201. **Harada, T. and Nikuni, Z.**, On a new bacterium which colors lignin red in urine. Part VI. Mechanism of the color-reaction, *J. Agric. Chem. Soc. Jpn.*, 23, 415, 1949.
202. **Harada, T.**, On a new bacterium in urine which makes lignin red. Part VII. The bacteriological characteristics of the L.R. bacterium, *J. Agric. Chem. Soc. Jpn.*, 24, 262, 1950.
203. **Harada, T.**, On a new bacterium in urine which makes lignin red. Part IX. Taxonomy and nomenclature of the bacterium, *J. Agric. Chem. Soc. Jpn.*, 28, 840, 1954.
204. **Barber, M., Brooksbank, B. W. L., and Kuper, S. W. A.**, Staphylococcal phosphatase, glucuronidase and sulphatase, *J. Path. Bact.*, 63, 57, 1951.
205. **Hare, R., Wildy, P., Billet, F. S., and Twort, D. N.**, The anaerobic cocci: gas formation, fermentation reactions, sensitivity to antibiotics and sulphonamides, classification, *J. Hyg.*, 50, 295, 1952.
206. **Young, L., Morrison, A. R., and Whitehead, J. E. M.**, Bacterial arylsulphatase, *Nature (London)*, 169, 711, 1952.
207. **Whitehead, J. E. M., Morrison, A. R., and Young, L.**, Bacterial arylsulphatase, *Biochem. J.*, 51, 585, 1952.
208. **Whitehead, J. E. M., Wildy, P., and Engbaëk, H. C.**, Arylsulphatase activity of mycobacteria, *J. Path. Bacteriol.*, 65, 451, 1953.
209. **Harada, T.**, On a new bacterium in urine which makes lignin red. Part VIII. Hydrolysis of urine-indican by microorganisms, *J. Agric. Chem. Soc. Jpn.*, 26, 95, 1952.
210. **Schultz-Haudt, S. D. and Scherp, H. W.**, Symbiotic production of phenolsulfatase by human gingival bacteria, *J. Bacteriol.*, 69, 665, 1955.
211. **Wayne, L. G., Juarez, W. J., and Nichols, E. G.**, Arylsulfatase activity of aerobic actinomycetales, *J. Bacteriol.*, 75, 367, 1958.
212. **Kubica, G. P. and Vestal, A. L.**, The arylsulfatase activity of acid-fast bacilli. I. Investigation of activity of stock cultures of acid-fast bacilli, *Am. Rev. Resp. Dis.*, 83, 728, 1961.
213. **Kubica, G. P. and Beam, R. E.**, The arylsulfatase activity of acid-fast bacilli. II. The differentiation of *Mycobacterium avium* from the unclassified Group III nonphotochromogenic mycobacteria, *Am. Rev. Resp. Dis.*, 83, 733, 1961.
214. **Kubica, G. P. and Rigdon, A. L.**, The arylsulfatase activity of acid-fast bacilli. III. Preliminary investigation of rapidly growing acid-fast bacilli, *Am. Rev. Resp. Dis.*, 83, 737, 1961.
215. **Steffen, D. L., Cocks, G. T., and Wilson, A. C.**, Microcomplement fixation in *Klebsiella* classification, *J. Bacteriol.*, 110, 803, 1972.
216. **Cocks, G. T. and Wilson, A. L.**, Enzyme evolution in the *Enterobacteriaceae, J. Bacteriol.*, 110, 793, 1972.

217. **Yamada, T., Murooka, Y., and Harada, T.,** Comparative immunological studies on arylsulfatase in bacteria of the family *Enterobacteriaceae:* occurrence of latent arylsulfatase protein regulated by sulfur compounds and tyramine, *J. Bacteriol.,* 133, 536, 1978.
218. **Harada, T. and Kono, K.,** Studies on phenolsulphatase from *Aerobacter aerogenes, J. Agric. Chem. Soc. Jpn.,* 28, 608, 1954.
219. **Harada, T. and Kono, K.,** An unidentified factor in the production of phenolsulfatase by *Enterobacteriaceae, Mem. Inst. Sci. Ind. Res., Osaka.,* 12, 183, 1955.
220. **Harada, T. and Hattori, F.,** Tyramine as an active factor for the production of phenolsulphatases by *Aerobacter aerogenes, Bull. Agric. Chem. Soc. Jpn.,* 20, 110, 1956.
221. **Fowler, L. R. and Rammler, D. H.,** Sulfur metabolism of *Aerobacter aerogenes,* II. The purification and some properties of a sulfatase, *Biochemistry,* 3, 230, 1964.
222. **Rammler, D. H., Grado, C., and Fowler, L. R.,** Sulfur metabolism of *Aerobacter aerogenes.* I. A repressible sulfatase, *Biochemistry,* 3, 224, 1964.
223. **Sukhumavasi, J., Murooka, Y., and Harada, T.,** Formation of sulfatases in *Klebsiella aerogenes, J. Ferment. Technol.,* 53, 62, 1975.
224. **Okamura, H., Yamada, T., Murooka, Y., and Harada, T.,** Purification and properties of arylsulfatase of *Klebsiella aerogenes:* identity of the enzymes formed by non-repressed and derepressed synthesis, *Agric. Biol. Chem.,* 40, 2071, 1976.
225. **Murooka, Y., Yim, M.-H., Yamada, T., and Harada, T.,** Affinity chromatography of *Klebsiella aerogenes* arylsulfatase on tyrosyl hexamethylenediamino-β1,3-glucan and immunoadsorbent, *Biochim. Biophys. Acta,* 485, 134, 1977.
226. **Murooka, Y., Yamada, T., Tanabe, S., and Harada, T.,** Immunological study of the regulation of cellular arylsulfatase synthesis in *Klebsiella aerogenes, J. Bacteriol.,* 132, 247, 1977.
227. **Arora, K. L., Dudani, A. T., and Krishnamurti, C. R.,** Survey of arylsulphatase activity in various pathogenic bacteria, *J. Sci. Ind. Res.,* 12, 502, 1953.
228. **Shrivastava, G. C., Arora, K. L. and Bhatanagar, S. S.,** Arylsulphatase activity in *Salmonella typhosa, Experientia,* 10, 493, 1954.
229. **Adachi, T., Murooka, Y., and Harada, T.,** Derepression of arylsulfatase synthesis in *Aerobacter aerogenes* by tyramine, *J. Bacteriol.,* 116, 19, 1973.
230. **Murooka, Y., Higashiura, T., and Harada, T.,** Genetic mapping of tyramine oxidase and arylsulfatase genes and their regulation in intergeneric hybrids of enteric bacteria, *J. Bacteriol.,* 136, 714, 1978.
231. **Henderson, M. J. and Milazzo, F. H.,** Arylsulfatase in *Salmonella typhimurium:* detection and influence of carbon source and tyramine on its synthesis, *J. Bacteriol.,* 139, 80, 1979.
232. **Stevens, R. L., Fluharty, A. L., Kilgrove, A. R., and Kihara, H.,** Microheterogeneity of arylsulphatase A from human tissue, *Biochim. Biophys. Acta,* 445, 661, 1976.
233. **Henderson, M. J.,** An investigation of arylsulfatase synthesized by *Salmonella typhimurium,* Ph.D. thesis, Queens' University, Kingston, Canada, 1979.
234. **Murooka, Y., Yim, M.-H., and Harada, T.,** Formation and purification of *Serratia marcescens* arylsulfatase, *Appl. Environ. Microbiol.,* 39, 812, 1980.
235. **Dodgson, K. S.,** Observations on the arylsulphatase of *Proteus vulgaris, Enzymologia,* 20, 301, 1959.
236. **Milazzo, F. H. and Fitzgerald, J. W.,** A study of arylsulfatase activity in *Proteus rettgeri, Can. J. Microbiol.,* 12, 735, 1966.
237. **Milazzo, F. H. and Fitzgerald, J. W.,** The effect of some cultural conditions on arylsulfatase of *Proteus rettgeri, Can. J. Microbiol.,* 13, 659, 1967.
238. **Fitzgerald, J. W. and Milazzo, F. H.,** Arylsulfatase multiplicity in *Proteus rettgeri, Can. J. Microbiol.,* 16, 1109, 1970.
239. **Fitzgerald, J. W. and Milazzo, F. H.,** Further studies on the heterogeneity of arylsulphatase from *Proteus rettgeri* and some properties of one chromatographic form of the enzyme, *Int. J. Biochem.,* 6, 769, 1975.
240. **Roy, A. B.,** The sulphatase of ox liver. IV. A note on the inhibition of sulphatases A and B, *Biochem. J.,* 59, 8, 1955.
241. **Dodgson, K. S. and Wynn, C. H.,** Studies on sulphatases. XIX. The purification and properties of arylsulphatase B of human liver, *Biochem. J.,* 68, 387, 1958.
242. **Dodgson, K. S. and Powell, G. M.,** Studies on sulphatases. XXVI. Arylsulphatase activity in the digestive juice and digestive gland of *Helix pomatia, Biochem. J.,* 73, 666, 1959.
243. **Dodgson, K. S., Melville, T. H., Spencer, B., and Williams, K.,** Studies on sulphatases. VIII. The arylsulphatase of a strain of *Alcaligenes metalcaligenes* isolated from intertidal mud, *Biochem. J.,* 58, 182, 1954.
244. **Dodgson, K. S., Spencer, B., and Williams, K.,** Studies on sulphatases. XI. The purification and properties of the arylsulphatase of *Alcaligenes metalcaligenes, Biochem. J.,* 61, 374, 1955.
245. **Shaw, I. S., Stein, W. D., and Wynn, C. H.,** Cyanide binding and inhibition of a purified arylsulphatase of *Alcaligenes metalcaligenes, Biochem. J.,* 112, 14p, 1969.

246. **Dodgson, K. S., Spencer, B., and Williams, K.**, Studies on sulphatases. XIII. The hydrolysis of substituted phenyl sulphates by the arylsulphatase of *Alcaligenes metalcaligenes*, *Biochem. J.*, 64, 216, 1956.
247. **Spencer, B.**, Studies on sulphatases. XX. Enzymic cleavage of aryl hydrogen sulphates in the presence of $H_2{}^{18}O$, *Biochem. J.*, 69, 155, 1958.
248. **Dodgson, K. S., Rose, F. A., and Tudball, N.**, Studies on sulphatases. XXIII. The enzymic desulphation of tyrosine *O*-sulphate, *Biochem. J.*, 71, 10, 1959.
249. **Dodgson, K. S., Spencer, B., and Williams, K.**, Examples of anticompetitive inhibition, *Nature (London)*, 177, 432, 1956.
250. **Chauncey, H. H., Lionetti, F., Winer, R. A., and Lisanti, V. F.**, Enzymes of the human saliva. I. The determination, distribution, and origin of whole saliva enzymes, *J. Dent. Res.*, 33, 321, 1954.
251. **D'Amato, R. F. and Eriquez, L. A.**, Rapid identification of *Neisseria gonorrhoeae* and *Neisseria meningitidis* by using enzyme profiles, *J. Clin. Microbiol.*, 7, 77, 1978.
252. **Harada, T. and Kamogawa, A.**, An example of application of the test paper method for detecting enzyme activities: isolation of different types of bacteria in arylsulfatase formation, *J. Ferment. Technol.*, 41, 132, 1963.
253. **Harada, T.**, The formation of sulphatases by *Pseudomonas aeruginosa*, *Biochim. Biophys. Acta*, 81, 193, 1964.
254. **Delisle, G. J. and Milazzo, F. H.**, The isolation of arylsulphatase isoenzymes from *Pseudomonas* aeruginosa, *Biochim. Biophys. Acta*, 212, 505, 1970.
255. **Delisle, G. J. and Milazzo, F. H.**, Characterization of arylsulfatase isoenzymes from *Pseudomonas aeruginosa*, *Can. J. Microbiol.*, 18, 561, 1972.
256. **Koster, J. F. and Veeger, C.**, The relation between temperature and inducible allosteric effects and the activation energies of amino acid oxidases, *Biochim. Biophys. Acta*, 167, 48, 1968.
257. **Baldwin, J. and Hochachka, P. W.**, Functional significance of isoenzymes in thermal acclimatization, *Biochem. J.*, 116, 883, 1970.
258. **Lehrer, G. M. and Barker, R.**, Conformational changes in rabbit muscle aldolase. Kinetic studies, *Biochemistry*, 9, 1533, 1970.
259. **Fitzgerald, J. W. and Payne, W. J.**, The regulation of arylsulphatase formation in *Pseudomonas* $C_{12}B$, *Microbios*, 6, 147, 1972.
260. **George, J. R. and Fitzgerald, J. W.**, Tyramine-mediated enhancement of bacterial arylsulphatase activity, *FEMS Microbiol. Lett.*, 3, 253, 1978.
261. **George, J. R.**, Tyramine-mediated enhancement of purified arylsulfatase from *Pseudomonas* $C_{12}B$, Ph.D. thesis, University of Georgia, Athens, 1979.
262. **George, J. R. and Fitzgerald, J. W.**, Tyramine-mediated enhancement of arylsulphatase purified from *Pseudomonas* $C_{12}B$, *Biochem. Soc. Trans.*, 7, 188, 1979.
263. **Fitzgerald, J. W. and George, J. R.**, Localization of arylsulphatase in *Pseudomonas* $C_{12}B$, *Appl. Environ. Microbiol.*, 34, 107, 1977.
264. **Fitzgerald, J. W. and Cline, M. E.**, The occurrence of an inducible arylsulphatase in *Comamonas terrigena*, *FEMS Microbiol. Lett.*, 2, 221, 1977.
265. **Maca, H. W. and Fitzgerald, J. W.**, Evidence for a periplasmic location in *Comamonas terrigena* of the inducible tyrosine sulfate sulfohydrolase, *Can. J. Microbiol.*, 25, 275, 1979.
266. **Fitzgerald, J. W., Maca, H. W., and Rose, F. A.**, Physiological factors regulating tyrosine-sulphate sulphohydrolase activity in *Comamonas terrigena:* occurrence of constitutive and inducible enzymes, *J. Gen. Microbiol.*, 111, 407, 1979.
267. **Maca, H. W.**, Physiological factors regulating tyrosine sulfate sulfohydrolase activity in *Comamonas terrigena*, M.S. thesis, University of Georgia, Athens, 1978.
268. **Selkon, J. B. and Mitchison, D. A.**, Atypical mycobacteria and drug-resistant tubercle bacilli isolated during a survey of untreated patients with pulmonary tuberculosis, *Tubercle*, 40, 141, 1959.
269. **Marks, J. and Trollope, D. R.**, A study of the anonymous mycobacteria, *Tubercle*, 41, 51, 1960.
270. **Jones, W. D., Abbott, V. D., Vestal, A. L., and Kubica, G. P.**, A hitherto undescribed group of nonchromogenic mycobacteria, *Am. Rev. Resp. Dis.*, 94, 790, 1966.
271. **Tsukamura, M.**, Identification of mycobacteria, *Tubercle*, 48, 311, 1967.
272. **Tsukamura, M.**, Numerical taxonomy of the genus *Nocardia*, *J. Gen. Microbiol.*, 56, 265, 1969.
273. **Tsukamura, M.**, Relationship between *Mycobacterium* and *Nocardia*, *Jpn. J. Microbiol.*, 14, 187, 1970.
274. **Gunnels, J. J. and Bates, J. H.**, Characterization and mycobacteriophage typing of *Mycobacterium xenopi*, *Am. Rev. Resp. Dis.*, 105, 388, 1972.
275. **Anneland, T. and Roland, M.**, Rapidly growing bacterium in chronic abscess. A case history, *Acta Path. Microbiol. Scand. Sect. B, Microbiol. Immunol.*, 80, 431, 1972.
276. **Gangadharam, P. R. and Hsu, K. H. K.**, *Mycobacterium abscessus* infection in a puncture wound, *Am. Rev. Resp. Dis.*, 106, 275, 1972.

277. **Tsukamura, M.**, Relationship between *Mycobacterium nonchromogenicum, Mycobacterium terrae, Mycobacterium novum* and subgroup "V" *(Mycobacterium triviale), Jpn. J. Microbiol.*, 15, 229, 1971.
278. **Yamamoto, K.-I. and Takahashi, Y.**, Simplification of the arylsulfatase test for the differential typing of mycobacteria, *Tuberc. Res.*, 32, 17, 1972.
279. **van Beneden, J.**, Enzymic equipment of mycobacteria, *Arch. Belg. Med. Soc., Hyg., Med. Trav. Med. Leg.*, 30, 1, 1972.
280. **Kubica, G. P.**, Differential identification of mycobacteria. VII. Key features for identification of clinically significant mycobacteria, *Am. Rev. Resp. Dis.*, 107, 9, 1973.
281. **Vestal, A. L.**, Procedures for the isolation and identification of mycobacteria, Publ. No. HSM 75-8230, U.S. Department of Health, Education and Welfare, Washington, D.C., 1975.
282. **Wayne, L. G.**, Recognition of *Mycobacterium fortuitum* by means of a three-day phenolphthalein sulfatase test, *Am. J. Clin. Path.*, 36, 185, 1961.
283. **Takeya, K., and Nakayama, Y., and Nakayama, H.**, Relationship between *Mycobacterium fortuitum* and *Mycobacterium runyonii, Am. Rev. Res. Dis.*, 96, 532, 1967.
284. **Binko, B. P. and Bartholomew, W. R.**, Enzymatic and immunological characterization of the *Mycobacterium fortuitum* complex, *J. Clin. Microbiol.*, 8, 246, 1978.
285. **Bartholomew, W. R.**, Sulfatase isoenzymes of mycobacteria, *Am. Soc. Microbiol. Bacteriol. Proc.*, 71, 87, 1971.
286. **Kubica, G. P., Baess, I., Gordon, R. E., Jenkins, P. A., Kwapinski, J. B. G., McDurmont, C., Pattyn, S. R., Saito, H., Silcox, V., Stanford, J. L., Takeya, K., and Tsukamura, M.**, A cooperative numerical analysis of rapidly growing mycobacteria, *J. Gen. Microbiol.*, 73, 55, 1972.
287. **Grange, J. M. and Clark, K.**, Use of umbelliferone derivatives in the study of enzyme activities of mycobacteria, *J. Clin. Path.*, 30, 151, 1977.
288. **Lien, T. and Schreiner, O.**, Purification of a derepressible arylsulfatase from *Chlamydomonas reinhardti*. Properties of the enzyme in intact cells and in purified state, *Biochim. Biophys. Acta*, 384, 168, 1975.
289. **Schreiner, O., Lien, T., and Knutsen, G.**, The capacity for arylsulfatase synthesis in synchronous and synchronized cultures of *Chlamydomonas reinhardti, Biochim. Biophys. Acta*, 384, 180, 1975.
290. **Agogbua, S. I. O. and Wynn, C. H.**, The application of affinity chromatography to the arylsulfatase B of human liver and to other arylsulfatases, *Biochem. Soc. Trans.*, 3, 405, 1975.
291. **Derappe, C. and Dubois, G.**, Preparation of a specific ligand for purification of arylsulfatase by affinity chromatography, *Experientia*, 32, 1354, 1976.
292. **van Etten, R. L. and Waheed, A.**, Subunit affinity chromatography and its application to the isolation of arylsulfatase A enzymes, *Arch. Biochem. Biophys.*, 202, 366, 1980.
293. **Chaudhury, S.**, Separation of arylsulfatase A and B in human leukocytes using Cibacron Blue Sepharose chromatography, *Sci. Cult.*, 44, 557, 1978.
294. **Ahmad, A., Surolia, A., and Bachhawat, B. K.**, Affinity chromatographic separation of arylsulfatase A and B using Cibacron Blue-Sepharose, *Biochim. Biophys. Acta*, 481, 542, 1977.
295. **Gelsema, W. J., DeLigny, C. L., and Klein Gebbink, J. A. G.**, Interaction of arylsulfatase A with SP-Sephadex. A note on the isolation of arylsulfatase A from human urine, *J. Chromatog.*, 169, 403, 1979.
296. **Fitzgerald, J. W.**, The occurrence of multiple electrophoretic forms of arylsulfatase in *Proteus rettgeri*, Ph.D. thesis, Queens' University, Kingston, Canada, 1968.
297. **Dodgson, K. S., Spencer, B., and Thomas, J.**, Studies on sulphatases. IX. The arylsulphatases of mammalian livers, *Biochem. J.*, 59, 29, 1955.
298. **Dodgson, K. S. and Rose, F. A.**, Sulphohydrolases, in *Metabolic Pathways*, Vol. 7, Greenberg, D. M., Ed., Academic Press, New York, 1975, chap. 9.
299. **Farooqui, A. A.**, Sulfatases, sulfate esters and their metabolic disorders, *Clin. Chim. Acta*, 100, 285, 1980.
300. **Fishbein, W. N.**, The structural basis for the catalytic complexity of urease; interacting and interconvertible molecular species (with a note on isoenzyme classes), *Ann. N.Y. Acad. Sci.*, 147, 857, 1969.
301. **Allen, E. and Roy, A. B.**, The sulphatase of ox liver. XI. The isoelectric focussing of a purified preparation of sulphatase B, *Biochim. Biophys. Acta*, 168, 243, 1968.
302. **Bleszynski, W. S. and Roy, A. B.**, Some properties of the sulphatase B of ox brain, *Biochim. Biophys. Acta*, 317, 164, 1973.
303. **Wortman, G.**, Arylsulfatase activity in beef and rabbit corneal extracts, *Arch. Biochem. Biophys.*, 97, 70, 1962.
304. **Chuaqui, C. A. and Lynn, K. R.**, Effects of ionization of *p*-nitrocatechol sulfate on its behaviour as a substrate for arylsulfatase, *Bio-Org. Chem.*, 8, 205, 1979.
305. **Fitzgerald, J. W.**, unpublished observations.

INDEX

A

E

G

H

I

J

K

L

M

N

Q

R

S

U

V

W

X

Y

Z

For Product Safety Concerns and Information please contact our EU
representative GPSR@taylorandfrancis.com
Taylor & Francis Verlag GmbH, Kaufingerstraße 24, 80331 München, Germany

www.ingramcontent.com/pod-product-compliance
Ingram Content Group UK Ltd.
Pitfield, Milton Keynes, MK11 3LW, UK
UKHW052052230926
475702UK00025B/1140
* 9 7 8 1 3 1 5 8 9 7 9 0 5 *